THE OZONE LAYER

THE OZONE LAYER

FROM DISCOVERY TO RECOVERY

↔ GUY P. BRASSEUR ↤

AMERICAN METEOROLOGICAL SOCIETY

Cover image: The Great Electriser of Martinus van Marum in the Oval Hall of Teylers Museum (Teylers Museum). This large electrostatic machine was developed circa 1785 by Dutch physicist Martinus van Marum to produce electric arcs. The peculiar "odor of electrical matter" produced by this electrical generator was later identified as the odor of ozone.

Published by the American Meteorological Society
45 Beacon Street, Boston, Massachusetts 02108

The mission of the American Meteorological Society is to advance the atmospheric and related sciences, technologies, applications, and services for the benefit of society. Founded in 1919, the AMS has a membership of more than 13,000 and represents the premier scientific and professional society serving the atmospheric and related sciences. Additional information regarding society activities and membership can be found at www.ametsoc.org.

Print ISBN: 978-1-944970-54-3
eISBN: 978-1-944970-55-0

Library of Congress Control Number: 2019920299

Contents

Preface

Without the existence of the protective atmospheric ozone layer that shields the biosphere from harmful solar ultraviolet radiation, life on Earth, as we know it, would not be sustainable. The high economic growth during the golden 1950s and 1960s gave little incentive to think about the consequences of the expanding industrial activities on the global environment, and to imagine a possible erosion of the ozone layer in response to the human enterprise. This optimistic view changed in the 1980s when the first signs of ozone depletion, following the release in the atmosphere of manufactured chemical products, were reported, particularly over the Antarctic continent. The public got alerted for the first time when, on November 7, 1985, the prominent science editor of the *New York Times*, Walter Sullivan, announced in his newspaper that "each October a *hole* appears in the ozone layer." By adopting this terminology suggested to him by atmospheric chemist Sherwood Rowland who was deeply concerned by the depletion of the ozone layer, Sullivan introduced a powerful environmental metaphor that translated the satellite images produced by NASA into a threatening reality and an influential message for the public and for the policy makers. The image was not entirely new. Already in 1934, the eminent British geophysicist Sydney Chapman in his presidential address to the Royal Meteorological

Society had suggested that, by injecting chemical agents in the atmosphere, an artificial ozone hole could be produced, which would allow astronomers to conduct optical observations of the Sun and the stars in a spectral range normally blocked by the atmosphere. But in the 1980s, the hole was not the result of a limited and deliberate action taken to satisfy the needs of a few astronomers. It was a continental-scale disturbance and an inadvertent consequence of our way of life. It required immediate action by all nations. The era of environmental diplomacy had begun.

The story of ozone started in 1839 as a mystery that lasted for almost 30 years. It was the detection of a pungent odor produced by the electrolysis of acidulated water in a laboratory of the University of Basel that triggered speculations about the nature of an unknown chemical compound released during the experiment. This gas repeatedly generated by electric discharges in air and called "ozone" as a reference to the Greek word "ozein" (to smell) was soon believed to be a modified form of oxygen with enhanced chemical affinities. In 1858, this "odorous oxygen" was found to be a permanent gas of the atmosphere, but it took almost ten additional years to unambiguously establish that ozone is a molecule composed of three oxygen atoms: OOO. Systematic investigations of the chemical and absorption properties of this gas attracted the attention of the best chemists and atmospheric scientists of the late nineteenth and the early twentieth centuries.

The history of the ozone research is fascinating. This book attempts to describe the evolution of knowledge during almost two centuries following the discovery of 1839. Written for a nonspecialized audience interested in the steady progress made by science, this volume is articulated around eleven chapters, which allows us to follow step by step the progress in the knowledge and the challenges encountered by the researchers of the time. The history of ozone research illustrates in many ways the methodologies adopted for scientific investigations in other fields of research with the intuition and doubts of individual scientists, their certainties sometimes denied a few years later, the vivid debates of ideas between scientific personalities and sometimes the disagreements and even severe conflicts between different protagonists. Research has always been a human endeavor influenced by the social and cultural environment in which scientists and engineers live and work. This volume shows that scientific breakthroughs by bright individuals, even if they were spectacular, were in most cases the result of small incremental steps based on earlier results. Early research was innovative probably because it was driven primarily by the curiosity of a few individual scholars (who did not have to justify their curiosity) rather than by large teams asked to develop goal-oriented and relevant projects with widespread appeal to funding agencies or review panels.

The first chapter highlights how early laboratory investigations led to the discovery of ozone in the middle of the nineteenth century, while Chapter 2 describes the first methods adopted to measure ozone in the atmosphere. Chapter 3 summarizes progress made on the determination of the spectroscopic properties of ozone and on the related instrumental development that led to the first detection of stratospheric ozone in 1912. Chapter 4 presents the first theoretical studies conducted around 1930 to explain the mechanisms that are responsible for the photochemical production and destruction of the molecule in the atmosphere. Chapter 5 discusses the first observational methods implemented to determine the vertical distribution of ozone in the upper atmosphere. The successive updates proposed after 1950 to complement the original ozone theory are the focus of Chapter 6. The questions related to the impact of human activities on the ozone layer are addressed in Chapters 7 and 8, and more specifically the expected effects of a planned fleet of supersonic aircraft in the 1970s, and of industrially manufactured chlorofluorocarbons emitted in the atmosphere since the 1950s. Chapter 9 highlights how the Antarctic ozone hole was discovered in the 1980s, and shows how this discovery led to intense discussions and even controversies. Finally, chapter 10 briefly reviews the processes identified in the second half of the twentieth century that are responsible for the formation of ozone pollution near the surface, particularly in urban areas. Chapter 11 presents some conclusions and highlights that the research that led to the protection of the ozone layer was indeed a successful story.

Acknowledgments

I would like to express my gratitude to several colleagues who have provided support to the production of this volume by reading the manuscript, providing material, and suggesting some changes or additions in the text. In particular, I would like to thank Rumen Bojkov, James Rodger Fleming, Idir Bouarar, Ian Galbally, Claire Granier, Daniel Marsh, Paul Newman, Brian Ridley, Ruan Xiao-xia, Susan Solomon, Johannes Staehelin, David Tarasick, Hans Volkert, Ying Xie and Christos Zerefos. Alexej Dobrynin, Ina Döge, Yong-Feng Ma and Yuting Wang are gratefully acknowledged for providing technical assistance. Thanks also to Sarah Jane Shangraw, managing editor at the American Meteorological Society, for her encouragements and support. Much of the book was written while working part-time at the Max Planck Institute for Meteorology in Hamburg, Germany, at the National Center for Atmospheric Research in Boulder, Colorado, USA, and at the Polytechnic University in Hong Kong. The National Center for Atmospheric Research is sponsored by the US National Science Foundation.

Introduction

O n May 16, 1985, the scientific journal *Nature* published a surprising article. The authors, Joseph C. Farman, Brian G. Gardiner, and Jonathan D. Shanklin of the British Antarctic Survey in Cambridge, UK, indicated that above the scientific base of Halley Bay in Antarctica, the amount of atmospheric ozone, measured systematically since the 1958 International Geophysical Year (see Box 0.1), began to decline dramatically during the southern spring. The effect had been particularly marked since the late 1970s.

This article produced a real shock among the members of the international scientific community. None of the mathematical models that calculate changes in the ozone layer had predicted such a trend. In addition, NASA, which is permanently monitoring the atmosphere through its satellites, had not announced any spectacular decrease in the amount of ozone in the southern polar regions. Only a researcher from Japan's meteorological services, Shigeru Chubachi, had indicated on a poster presented at the 1984 Quadrennial Ozone Conference in Halkidiki, Greece, that ozone levels measured at the Japanese base in Syowa, Antarctica, had been abnormally low in September 1982 and 1983. Interestingly, no similar ozone layer disturbance had been reported anywhere else in the world.

Box 0.1. The International Geophysical Year 1957 to 1958

On April 5, 1950, during a dinner hosted by James Van Allen, a professor at The John Hopkins University in Baltimore, Maryland, and by his wife Abigail, a few prominent scientists (Lloyd Berkner, Sydney Chapman, S. Fred Singer, and Harry Vestine) suggested that the time was ripe to organize a Geophysical Year with strong international participation in order to investigate the complex and interrelated processes that affect planet Earth. This audacious idea was presented in different circles a few weeks later: first at a meeting held at the California Institute of Technology (Caltech) in Pasadena, California, on the upper atmosphere with the presence of about 20 geophysicists (including Belgian aeronomer Marcel Nicolet), and then in July 1950 at a conference on the physics of the ionosphere held at The Pennsylvania State University. With the strong support of American scientists, the project was then submitted to the International Council of Scientific Unions and was adopted in 1952 by this organization. A steering committee, established under the French name of "Comité Spécial de l'Année Géophysique Internationale" (CSAGI) (see the figure below), was charged to implement the activities of what became "The 1957–1958 International Geophysical Year" (IGY).

Figure 0.1. Meeting of the "Comité Spécial de l'Année Géophysique Internationale" (CSAGI) (Special Committee of the International Geophysical Year [IGY]) in Brussels, Belgium, with left to right Vladimir Belousov (Soviet Union), Lloyd V. Berkner (United States, vice chair), Marcel Nicolet (Belgium, secretary general), Jean Coulomb (France), and Sydney Chapman (United Kingdom, chair). Photograph from Life Magazine.

Twenty-six countries initially decided to join the program and to conduct coordinated observations. However, with the political tensions produced by the Cold War, scientific communication between the western and the eastern blocs had been almost entirely interrupted, and hence the Soviet Union postponed its participation. It decided to join the program only in October 1954 after the death of Joseph Stalin in the previous year. China refused to participate as long as the Republic of China (Taiwan) remained a member of the IGY.

Several interesting initiatives were undertaken. New ozone stations, for example, were installed in different parts of the world including Antarctica. World Data Centers providing geophysical information, including measured ozone concentrations, were established. The space race between the Soviet Union and the United States started: the first artificial Earth-orbiting satellite *Sputnik 1* was launched on October 4, 1957 by the Soviet Union and the first US satellite *Explorer 1* was put in the orbit on January 31, 1958. Continuous monitoring of carbon dioxide (CO_2) was initiated at the station of Mauna Loa in Hawaii, and the famous Van Allen radiation belts were discovered by *Explorer 1*.

The IGY can be viewed as the largest coordinated geophysical research program ever implemented by the international community.

The announcement of the formation of a real "hole" in the ozone layer was propagated by the press of all the countries. For many people, this represented a real ecological disaster whose magnitude and future consequences for life on Earth were apprehended.

The ozone layer, located mainly in the stratosphere, the region of the atmosphere extending from approximately 10 to 50 km altitude (see Box 0.2), is known to protect living beings from ultraviolet radiation emitted by the sun. This radiation has harmful effects on living cells and in particular destroys DNA. Without the presence of ozone in the atmosphere, which acts as a protective shield against ultraviolet radiation, radiation, life in its present form would not have been possible. In particular, it is known that even a partial destruction of the ozone layer would lead to a considerable increase in the number of skin cancers among the human population.

Two questions therefore immediately arose: What are the mechanisms responsible for the formation in Antarctica—and only in this region—of a hole in the ozone layer whose size (30 million km^2) corresponds to nearly

Box 0.2. The stratosphere

Exploring the upper layers of the atmosphere was made possible at the end of the nineteenth century with the advent of balloons and the existence of devices capable of measuring air temperature and pressure. These investigative techniques enabled Léon Philippe Teisserenc de Bort (1855–1913), who worked at the Trappes Meteorological Observatory, and Richard Assmann (1845–1918), who became Director of the Royal Aeronautical Observatory of Prussia at Lindenberg (Figure 0.1) in 1905, to show for the first time that the decrease in temperature with altitude stopped at around 11 km. Above this level, the temperature is first constant with altitude and then rises to a maximum at 50 km above sea level. This region of the atmosphere is characterized by vertical air movements of low amplitude, isolating the different levels of altitude from each other. This layer, vertically stratified, was called "stratosphere" by Teisserenc de Bort, from the Latin "stratum". It extends up to 50 km in altitude (the upper limit of the stratosphere called the stratopause).

Figure 0.2. (Left) French Meteorologist Léon Teisserenc de Bort. Source: https://en.wikipedia.org/wiki/L%C3%A9on_Teisserenc_de_Bort. (Right) Prussian scientist Richard Assmann. Source: Stadtarchiv Magdeburg (Fotographie); Meteorologisches Observatorium Lindenberg (Ölgemälde); Das Wetter. Sonderheft für R. A., 1915. Both scientists discovered the stratosphere almost simultaneously and independently.

50 times the surface area of France? Is it a natural disturbance that has repeated itself several times in the past, or is it a new phenomenon that has resulted from human activity? The answers to these questions were rapidly provided by the scientific community. It is indeed a disturbance resulting from human activities, it results from the emission in the atmosphere of products called chlorofluorocarbons (CFCs) and generated by industry and by a large number of domestic and industrial applications.

The history of ozone began in a university laboratory in Basel, Switzerland, where a chemistry professor, Christian Friedrich Schönbein, detected a peculiar odor while conducting an experiment on water electrolysis. The cause of this smell was not identified until about 20 years later, when it was established that ozone is an allotrope[1] form of oxygen. The chemical symbol of this molecule is O_3. The discovery of this substance caught the attention of the scientists of the time and, in a letter to Schönbein on March 12, 1847, chemist J. Berzelius wrote that this was one of the most beautiful discoveries ever made.[2] A few years later, it was noticed that the ozone molecule absorbs ultraviolet radiation very effectively, and that its presence in the atmosphere is therefore an essential element for the maintenance of life on Earth.

The discovery in 1858 by the French agronomist Jean-Auguste Houzeau that ozone is a permanent gas of the atmosphere opened up a new field of research. The first step was to establish accurate techniques for measuring the concentration of this gas in the air. In the early part of the 20th century, the vertical distribution of this gas was determined after it was recognized that ozone had to be more abundant in the upper atmosphere than near the Earth's surface. In the 1920s and 1930s, prominent scientists including Charles Fabry in France, Gordon Dobson in the United Kingdom, Paul Götz in Switzerland, and Erich and Victor Regener in Germany played a key role in observing ozone and, after developing sophisticated techniques, discovered the presence of an ozone layer in the stratosphere, that is, in the region of the atmosphere extending between 15 and 50 km above sea level (see Box 0.2). This ozone layer, whose essential property is to absorb a large fraction of ultraviolet solar radiation, is characterized by a maximum concentration located at an altitude of about 25 km (Figure 0.3). Ozone is also

---

1. Term proposed by the Swedish chemist Jöns Jacob Berzelius to define several forms of the same chemical element within the same phase.

2. "Ihre Endeckung von Ozon ist aus diesen Gesichtpunkte eine der schönster die je gemacht worden sind."

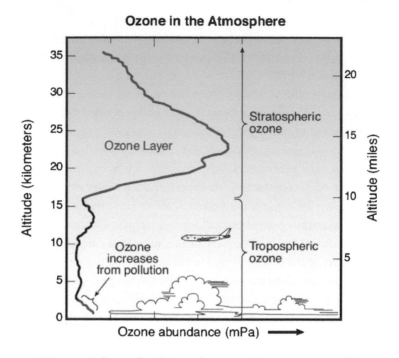

Figure 0.3. Mean vertical mean distribution of ozone partial pressure expressed here in milli-Pascals (mPa), which shows the maximum concentration in the stratosphere at about 25 km altitude and the lowest abundance of ozone in the troposphere with higher amounts due to pollutant emissions at the Earth's surface. Source: Fahey, D. W. and M. I. Hegglin (coordinating lead authors). *Twenty Questions and Answers About the Ozone Layer: 2010 Update, Scientific Assessment of Ozone Depletion* (Geneva, Switzerland: World Meteorological Organization, 2011), 72 pp.

present, but in smaller amounts, in the troposphere,[3] the atmospheric layer that extends from the Earth's surface to the tropopause,[4] the boundary with

---

3. The troposphere is the layer of the atmosphere that extends from the ground up to altitudes of about 18 km in the tropics, 12 km at mid-latitudes and 6 to 8 km in the polar regions. It contains 75% of the air mass and is subject to weather disturbances.

4. According to the World Meteorological Organization, the tropopause is defined as "the lowest level at which the lapse rate decreases to 2 °C/km or less, provided that the average lapse rate between this level and all higher levels within 2 km does not exceed 2 °C/km". Other definitions, often preferred by atmospheric chemists, are based on the rapid change with height (discontinuities) in atmospheric static stability or in the value of stratospheric tracers including ozone concentration and potential vorticity.

the lower stratosphere. Tropospheric ozone, through a series of complex chemical reactions, determines the "oxidizing potential of the atmosphere," that is, its ability to oxidize and thus destroy certain chemical species emitted at the Earth's surface, including most chemical pollutants. Finally, near urban centers and in industrial areas, ground-level ozone can be produced in large quantities by complex chemical processes that involve air pollutants, including nitrogen oxides and hydrocarbons. Ozone pollution, observed mainly during the summer in stagnant air exposed to intense sunlight, produces harmful effects on human health and, in particular, causes respiratory diseases and cardiac problems. Ozone is a powerful oxidant that irritates the respiratory tract. It also inhibits plant growth and therefore reduces the yield of agricultural crops with large economic consequences.

The first atmospheric measurements of ozone were complemented by theoretical studies that attempted to explain the origin and fate of this chemical species. The first theory, proposed in 1929 by the British geophysicist Sydney Chapman, describes the processes of ozone formation and destruction by invoking only five chemical reactions involving only oxygen species. It was completed in the 1950s by adding the effects of water vapor and hydrogenated radicals, and in the 1970s by recognizing the important role of nitrogen and halogenated compounds (chlorine and bromine). The latter work highlighted the vulnerability of ozone to atmospheric releases of chemicals produced by human activity. The formation of the ozone hole at the end of the twentieth century demonstrated the urgent need to safeguard the global environment by removing from the atmosphere industrially manufactured products that are likely to alter the ozone layer. From that moment onwards, the ozone issue was no longer only seen as a purely academic problem, but became a societal issue in the same way as global warming.

In this book, we describe, sometimes with interesting anecdotes and often through the discovery of key personalities, the development of new scientific concepts that have shaped ozone research since the discovery of this gas in the mid-nineteenth century. The results of this research provided the knowledge at the end of the twentieth century that led to international agreements to protect humanity from the destruction of this gas in the stratosphere. This historical journey will allow us to follow the evolution of our scientific knowledge on stratospheric chemistry and dynamics, and more recently on air quality issues, for nearly two centuries. In particular, we will demonstrate the important role of fundamental research in solving a global problem affecting the earth's environment, climate, and human health. The history of ozone research is marked by events that are typical of the scientific method and is a perfect illustration of how knowledge is progressing. We will present hypotheses that were often contested and then eventually accepted or rejected. We will show how intense debates and disagreements between scientists were settled by the information brought by laboratory or field experiments. We will also show how "truths" that were believed to be universal and permanent were sometimes called into question. We will highlight how new observations took researchers by surprise and led to new investigations and new research programs. And finally, we will point the reader to the difficulty of convincing decision makers of research results that do not necessarily correspond to their views or to their interests.

CHAPTER ONE

The First Steps

The Discovery of Ozone

Christian Friedrich Schönbein (1799–1868), a prominent professor of chemistry at the University of Basel (Figure 1.1), was a specialist in electrolysis. He invented the fuel cell and discovered the explosive properties of nitrocellulose. Passionate about chemistry already at a young age, he started working as an apprentice at a pharmaceutical company in Böblingen, Germany, at the age of 13. In order to satisfy his ambition to explore the complex processes of chemistry, he passed several examinations organized by a professor of chemistry in Tübingen. In 1820, he did an apprenticeship training in a textile dyer company before enrolling as a student at the universities of Erlangen and Tübingen. In 1826, he moved to England to become a teacher in a reformatory house and later completed his studies in physics and chemistry at the Sorbonne in Paris. He joined the University of Basel in 1828, where he was promoted to "Ordinary Professor" in 1835. Schönbein became one of Europe's most renowned chemists while being involved in the political affairs of the city of Basel. The famous chemist Justus von Liebig (1803–1873), professor at the University of Giessen and later of Munich, one of the founders of industrial agronomy, who met Schönbein for the first time in 1853, seemed to appreciate the company of his Basel colleague as in a letter

Figure 1.1. Christian Friedrich Schönbein, professor at the University of Basel. Source: Photogravure after a painting by Beltz, c. 1860; original photo from the University of Basel Library.

to Friedrich Wöhler (1800–1882), professor at the University of Göttingen,[1] he wrote:

> Schönbein's sense of humor is priceless; if only I had his stomach ...[2]

During an experiment on the electrolysis of acidic water conducted in early 1839, Schönbein noticed a peculiar odor, similar to that produced by the oxidation of white phosphorus. It was very similar to the smell of "electric material" that Martin van Marum[3] (1750–1837), a physician in the city of Harlem in the Netherlands, had reported 40 years earlier and that accompanied

---

1. Friedrich Wöhler was the first to synthesize urea.

2. "Schönbein's Humor ist unschätzbar; wenn ich nur seinen Magen hätte."

3. Martin(us) van Marum, born in Delft in the Netherlands, studied medicine and philosophy in Groningen. He developed a large electrostatic generator that he used to study chemical processes. Being very much influenced by the work of the French chemist Antoine Lavoisier (1743–1794), he introduced modern chemistry in the Netherlands. In 1784, he became the director of the Teylers' museum in Harlem and ten years later, the Director of the Dutch Society of Science. He was a member of several academic societies.

Figure 1.2. (Left) Arthur de la Rive. Source: Wikipedia. (Right) Jean Charles Galissard de Marignac. Source: Orden pour le mérite für Wissenschaften und Künste. These two scientists showed that ozone is produced in an enclosure containing only oxygen.

the production of electric arcs by a friction electrostatic generator that the Dutch scientist had built. In fact, as highlighted in the *Odyssey* and *Iliad* by Greek poet Homer (born sometimes between the twelfth and eighth century BC), the "divine odour of Zeus" had already been observed during Antiquity in relation to "Jupiter's thunders" (lightning flashes).

On March 13, 1839, Schönbein went to a meeting of the Natural Sciences Society (Naturforschung Gesellschaft) in Basel and made his discovery public; the minutes of the meeting state, "Professor Schönbein told the Society of an odour that develops during the electrolysis of water."[4]

It was only a few months later, however, in 1840, during a lecture he gave at the Academy of Sciences in Bavaria, that Schönbein suggested that the odor produced during the electrolysis of water should be attributed to a chemical substance, but he had not yet been able to identify this substance. Was it a particular molecule or a mixture of several chemical elements? Or simply, as suggested by the professor of Physics of Geneva, Arthur de la Rive (1801–1873; Figure 1.2), an odor produced by fine particulates released from the material constituting the electrode of the electrolysis cell? Schönbein

---

4. "Herr Prof. Schönbein macht die Gessellschaft [...] aufmerksam dass bei der Electrolyse des Wassers ein Geruch entwickelt wird."

> **Box 1.1. The characteristic smell of ozone is observed near thunderstorms**
>
> M. Buchwalder, a Swiss engineer, was one day on the summit of the Senlis, near Appenzell, with his servant. They were reclining in a little tent erected in the snow, when they were suddenly enveloped in a thunderstorm. The servant was killed on the spot, and, immediately after, the tent was filled with a very powerful and remarkable odor. One day when M. Schönbein was experimenting with ozone, his laboratory being filled with its odor, he received a visit from M. Buchwalder, who recognized the odor as precisely similar to that which he had perceived in his tent on the mountain.
>
> Incident narrated by C. Schönbein and reproduced from the text published in a book by C. Fox entitled *Ozone and Antozone* with reference to an article published in the Union Médicale, p. 82, 1853.

refutes this last hypothesis by arguing that the same odor, detected in the vicinity of thunderstorms, was probably due to lightning strokes. He therefore argued that the odor recorded in his laboratory must be linked to the electrical arc produced during his experiment (see Box 1.1).

During a lunch at the university, Schönbein shared his discovery with his colleague Wilhelm Vischer-Bilfinger (1808–1874), a philologist and specialist in ancient Greek literature. Vischer suggested to name this odorous property "*ozon*" after the Greek word όζειν (ozein, to smell).[5] Schönbein adopted this idea, which he communicated in 1840 to French academician François Arago[6]

---

5. The name *ozon* chosen by Schönbein was not unanimously accepted. Anglophones adopted the term "*ozone*" after adding a silent *e* to the original German name, but until 1870, French researchers preferred to talk about "*oxygène odorant*" (odorous oxygen) or "*oxygène naissant*" (nascent oxygen), sometimes "*oxygène ozonisé*" (ozonized oxygen). Other names were "*electricized oxygen*," "*allotropic oxygen*," or in German "*erregter Sauerstoff*" (excited oxygen). Finally, the name of ozone was universally adopted. Nineteenth-century French-language authors often write *ozône* with a circumflex accent on the letter *o*.

6. Dominique François Jean Arago, known as François Arago, was a French mathematician, physicist, astronomer, and a republican politician. After his studies at Ecole Polytechnique in Paris, he worked at Bureau des Longitudes and at the Paris Observatory. He was elected a member of the Academy of Sciences at the age of 23. He was member of the French House of Representatives (Chambre des Députés) from 1831 to 1848 representing his region of the eastern Pyrenees, and for a short

(1786–1853) in a letter on the "nature of the odor that manifests itself in certain chemical actions." This letter was read at the meeting of the Academy of Sciences in Paris on April 27, 1840. In an article published in German on April 8, 1840, and after having exposed the results of his work, Schönbein wrote,

> After these remarks, I still make the proposal to name *ozon* the odorant principle if future research shows that it behaves as an elemental or composed halogen.[7]

A similar letter in which he mentioned the presence of the new odorous substance "accompanying electricity" was sent to his friend Michael Faraday (1791–1867) and read at a session of the Royal Society in London.

Identification of the Ozone Molecule

The nature of the "electric smell" was hotly debated because it was highly controversial. Schönbein was initially convinced that the electric discharge produced during the electrolysis of the water produces a gaseous substance that he first regarded as a halogen (chlorine or bromine). In his letter to Arago, he wrote,

> Since I am pretty sure that the odorant principle must be classified according to the kind of body to which chlorine and bromine belong, i.e. elemental and halogenated substances, I propose to call it ozone. As I am convinced that this body is always released into the air in fairly noticeable quantities, when the weather is stormy, I propose to do a series of experiments this year to highlight the presence of ozone in our atmosphere.[8]

---

period of time in 1848 was Minister of War and Minister of the Navy. He became President of the Executive Power Commission on May 11, 1848 and served in this capacity as provisional head of state until June 24, 1848.

7. "Aus diese Bemerkung, knüpfe ich noch den Vorschlag das riechende Prinzip *Ozon* zu nennen, wenn es sich bei ferneren Untersuchungen entweder als elementarer oder zusammengesetzter Salzbildner verhalten sollte."

8. "Etant à peu près sûr que le principe odorant doit être classé au genre de corps auquel appartient le chlore et le brome, c'est-à-dire les substances élémentaires et halogènes, je propose de lui donner le nom de ozone. Comme je suis convaincu que ce corps se dégage toujours dans l'air en quantité assez notable, lorsque le temps est

Yet seven years later, he revised himself and, in an article published in 1847, concluded that the substance emitted should contain oxygen and hydrogen atoms, and that it would probably be a high-order oxide HO_3 rather than the hydrogen peroxide H_2O_2, a chemical substance discovered by Louis Jacques Thénard[9] (1777–1857). Later, Schönbein suggested that ozone was actually the peroxy radical HO_2. De la Rive was of a different opinion based on the experiments he had carried out in his laboratory and of which he reported in a letter to Arago in 1845:

> We passed through a tube a stream of oxygen perfectly pure and perfectly dried up [...] Thus [...] ozone comes only from oxygen and, in order to have the manifestation of it, the simplest and most direct way is to produce a succession of electric sparks through a volume of pure oxygen.[10]

Encouraged by these experimental results, Jean Charles Galissard de Marignac (1817–1894; Figure 1.2), professor of chemistry at the Geneva Academy, who had also found that the characteristic smell of ozone appeared when an electric discharge was produced in a chamber containing pure oxygen, suggested that ozone is an allotropic form of oxygen. This conclusion was immediately rejected by Schönbein who considered that the enclosure used by de Marignac probably contained impurities and in particular traces of water vapor. In an article, "On the Nature and Production of the Ozone," published in the Electricity Archives in 1845, de Marignac wrote,

> The attention of chemists and physicists has been excited to a high degree by the remarkable phenomena which have been studied by Mr. Schönbein. The interest in the curious properties of ozone was further enhanced by the ingenious hypothesis through which the learned professor of Basel believed that he could explain these properties. However, chemists cannot easily adopt

---

orageux, je me propose de faire une série d'expériences cette année pour mettre en évidence la présence d'ozone dans notre atmosphère."

9. Louis Jacques Thénard was a French Chemist, professor at Ecole Polytechnique in Paris and member of the French Academy of Sciences. He worked closely with Joseph Louis Gay-Lussac (1778–1850).

10. "Nous avons fait passer à travers un tube un courant d'oxygène parfaitement pur et parfaitement desséché [...] Ainsi, [...] l'ozone ne provient que de l'oxygène et, pour en avoir la manifestation, le moyen le plus simple et le plus direct, c'est de faire passer à travers l'oxygène une succession d'étincelles électriques."

a theory that overturns all their doctrines on the simple nature of nitrogen, without collecting fully convincing evidence. [...] When the air is subject to a series of electric discharges [...], ozone develops, without the presence of water vapor being necessary. It seems to us from these facts that the oxygen is likely, in certain circumstances, to undergo a particular modification which enhances its chemical affinities, and makes it able to combine directly with bodies on which it is without action in its ordinary state. Ozone is nothing else than oxygen in this particular state of chemical activity, which results from the influence of an electric current.[11]

Until 1850, Schönbein remained convinced that ozone was indeed a compound of oxygen and hydrogen. But with the laboratory measurements made in Geneva, he finally had to realize that ozone contains only oxygen atoms. But in what form?

Two theories confronted each other immediately. The first theory expressed by Schönbein was based on the fact that the amount of ozone produced by an electric discharge in an oxygen-containing enclosure is unexpectedly low. It implies therefore that a chemical destruction process involving an unidentified substance reduces the net formation rate of ozone. In 1858, Schönbein imagined that ozone was a negatively electrified form of oxygen, and that its abundance was balanced by the presence of a similar quantity of positively electrified oxygen. In a letter to Michael Faraday, he used the name *antozone* to describe the positively charged oxygen substance, and considered that the "ordinary" oxygen molecule (O_2) resulted from the

---

11. "L'attention des chimistes et des physiciens a été excitée à un haut degré par les phénomènes remarquables dont l'étude a été faite par Mr. Schönbein. L'intérêt qu'inspirent les propriétés curieuses de l'*ozône*, a été augmenté encore par l'ingénieuse hypothèse à l'aide de laquelle le savant professeur de Bâle a cru pouvoir expliquer ces propriétés. Toutefois les chimistes ne pouvaient pas facilement adopter une théorie qui renverserait toutes leurs doctrines sur la nature simple de l'azote, sans s'appuyer sur des preuves entièrement convaincantes. [...] Quand l'air est soumis à une série de décharges électriques [...], l'*ozône* se développe, sans que la présence de vapeur d'eau soit nécessaire. [...] L'*ozône* de développe avec la plus grande facilité dans l'*oxigène* parfaitement pur et sec [...] Il nous semble de ces faits que l'*oxigène* est susceptible, dans certaines circonstances, de subir une modification particulière qui exalte ses affinités chimiques, et le rend apte à se combiner directement avec des corps sur lesquels il est sans action dans son état ordinaire. L'*ozône* n'est pas autre chose que l'*oxigène* dans cet état particulier d'activité chimique, qui lui est imprimé par l'influence d'un courant électrique."

Box 1.2. Letter from Michael Faraday to Christian Schönbein highlighting the disarray in which he finds himself after hearing the different theories presented successively by Schönbein

Royal Institution Nov. 13, 1858
My Dear Schönbein,

Daily and hourly am I thinking about you and yours, and yet with as unsatisfactory a result as it is possible for me to have. I think about Ozone, about Antozone, [...] and it all ends in a giddiness and confusion of the points that ought to be remembered.

I want to tell our audience what your last results are upon this most beautiful investigation, and yet am terrified at the thoughts of trying to do so, from the difficulty of remembering from the reading of one letter to that of another, what the facts in the former were. I have never before felt so seriously the evil of loss of memory and of clearness in the head; and though I expect to fail some day at the lecture table, as I get older, I should not like to fail in ozone, or in any thing about you.

M. Faraday

combination of ozone and antozone. The contradictory hypotheses that Schönbein made over the years to characterize "the odor of electricity" puzzled Faraday who underlined his deep confusion in a letter sent to Schönbein in 1858 (see Box 1.2). At this point, it remained to characterize the chemical nature and the properties of antozone and to first isolate this compound in the laboratory.

Georg C. F. Meissner (1829–1905), professor of Physiology at Göttingen University (Figure 1.3), expressed interested in the problem. In laboratory experiments carried out in 1863, he found that, if electrified oxygen circulates in an enclosure containing moist air, fog was produced. Meissner first named this unidentified substance *atmizone* (from the Greek, ατμιζω, emitting smoke), but he quickly recognized that it was identical to Schönbein's antozone. Antozone soon received a lot of attention because it was believed to be associated with smoke produced by cigarettes, chimneys, and even gunpowder.

The second theory was in line with the work performed by de la Rive and de Marignac. These two researchers had shown that the characteristic smell of ozone occurs also when the "electrified" gas is composed of pure oxygen.

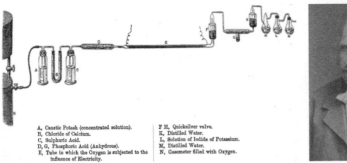

A, Caustic Potash (concentrated solution).
B, Chloride of Calcium.
C, Sulphuric Acid.
D, G, Phosphoric Acid (Anhydrous).
E, Tube in which the Oxygen is subjected to the influence of Electricity.

F H, Quicksilver valve.
K, Distilled Water.
L, Solution of Iodide of Potassium.
M, Distilled Water.
N, Gasometer filled with Oxygen.

Figure 1.3. The device (left, reproduced from Fox, 1873) designed by Georg Meissner (right; source: National Library of Medicine) to produce antozone (https://pixels.com/featured/georg-meissner-national-library-of-medicine.html).

Interestingly, in 1848, Canada's Geological Survey chemist Thomas Sperry Hunt (1826–1882), using an analogy with the molecule of sulfur dioxide (SO_2), stated that a molecule composed of three oxygen atoms must exist, but he did not identify this molecule as being ozone. Ten years later, in 1858, the famous Prussian physicist Rudolph Clausius (1822–1888) suggested that ozone is probably nothing more than a single oxygen atom (O).

Real progress toward the identification of the chemical nature of ozone was made in 1857 when Irish chemistry professor Thomas Andrews (1831–1885) and Scottish mathematician Peter Guthrie Tait (1831–1901) made densitometry measurements at Queens University in Belfast, and showed that ozone is denser than oxygen. This led William Odling (1829–1921), a lecturer at St Bartholomew Hospital Medical School in London, to state in his "Manual of Chemistry" published in 1861 that the ozone consists of three oxygen atoms. However, it was finally the Swiss chemist and physicist from Geneva, Jacques-Louis Soret (1827–1890), former student of Marignac (Figures 1.4 and 1.5), who accurately determined the density of ozone. In a note published in 1863 on "The volumetric relationships of ozone," he wrote:

A large number of chemists and physicists admit that the oxygen molecule is [...] formed from the combination of 2 atoms and constitutes an oxygen oxide OO.[12]

---

12. "Un grand nombre de chimistes et physiciens admettent que la molécule d'oxygène est [...] formée de la réunion de 2 atomes et constitue un oxyde d'oxygène OO."

Figure 1.4. Jacques-Louis Soret showed that ozone is a molecule made up of three oxygen atoms. Reproduced from Wikipedia https://en.wikipedia.org/wiki/Jacques-Louis_Soret.

And by adopting a similar view for ozone, he added that

> if ozone is an allotrope state of oxygen, one is led to suppose that the ozone molecule results from an atomic arrangement [...]. For example, it could be conceivable that an ozone molecule was composed of three OOO atoms and constitutes an oxygen dioxide.[13]

Soret accurately measured the ozone density by using an ingenious method based on Graham's diffusion law.[14] He confirmed his results in 1868, shortly

---

13. "Si l'ozone est un état allotrope de l'oxygène, on est amené à supposer que la molécule d'ozone résulte d'un arrangement atomique [...]. On pourrait, par exemple, concevoir qu'une molécule d'ozone fût composée de trois atomes OOO et constituât un bioxyde d'oxygène."

14. The Scottish chemist Thomas Graham (1805–1869) demonstrated in 1833 that the rate of diffusion of a gas is inversely proportional to the square root of its molar mass.

CHIME. — *Recherches sur la densité de l'ozone.* Note de **M. J.-L. Soret**, présentée par M. Regnault.

« On peut résumer de la manière suivante ce que l'on sait actuellement sur les relations volumétriques de l'ozone (1) :

» 1° L'oxygène ordinaire diminue de volume lorsqu'on l'ozonise, c'est-à-dire lorsqu'on en convertit une partie en ozone, par exemple en l'électrisant.

» 2° Lorsqu'on traite de l'oxygène chargé d'ozone par l'iodure de potassium et d'autres corps oxydables, l'ozone disparaît sans que l'on observe de changement dans le volume du gaz.

» 3° Sous l'action de la chaleur, l'oxygène chargé d'ozone subit une expansion égale au volume qu'occuperait la quantité d'oxygène que le gaz aurait été susceptible d'abandonner à l'iodure de potassium.

» Les faits conduisent à supposer que l'ozone est un état allotropique de l'oxygène consistant en un groupement moléculaire de plusieurs atomes de ce corps. L'une des hypothèses les plus simples à cet égard est celle que j'ai indiquée précédemment (2) et dans laquelle on considère la molécule d'oxygène ordinaire comme formée de 2 atomes OO, et la molécule d'ozone comme formée de 3 atomes OO,O. Alors l'ozone contiendrait son volume d'oxygène ordinaire; traité par l'iodure de potassium, il perdrait 1 atome O, sans changement de volume; décomposé par la chaleur, il subirait une expansion de la moitié de son volume. Sa densité théorique devrait être dans ce cas $1\frac{1}{2}$ fois celle de l'oxygène, soit 1,658.

Figure 1.5. Excerpt from an article published by J. L. Soret in which the author suggests that ozone is an allotropic form of oxygen. Source: J. L. Soret, Comptes Rendus hebdomadaires, *Acad. Sci.*, *61*, 941–944, 1865.

before Schönbein's sudden death in Baden-Baden on August 29, 1868, while the Basel Professor was traveling to Bad Wilbad in Württemberg to be treated against gout. Soret's measurements thus put an end to the controversy that had troubled the world of chemists for nearly 30 years: ozone is a molecule composed of three oxygen atoms.

One question remained. What was this antozone that Meissner had observed (Box 1.3)? Already in 1863, Lambert Heinrich von Babo (1818–1840), professor of chemistry at the medical faculty at the University of Freiburg claimed that antozone was in fact hydrogen peroxide (H_2O_2). This idea was supported in 1866 by Carl Weltzien (1813–1840), a professor at the Technische Hochschule of Karlsruhe. And, based on their laboratory

Box 1.3. Ozohydrogen, antozone and oxozone: A wild goat chase

The first studies on the nature and properties of ozone faced major difficulties. In 1785, van Marum had concluded from his studies in the Netherlands that the smell he had detected was nothing but a manifestation of the "electrical matter" since this phenomenon occurred by repeated action of electric spark discharges. By the work of Schönbein, this concept was replaced in 1840, by that of a "chemical substance" produced during the electrolysis of water, a substance that the chemist did not manage to identify completely.

Other studies that followed the initial discoveries produced some confusion and opened the way to false trails. These are well described by Mordecai B. Rubin in a series of articles on "the History of the Ozone." For example, in Würzburg around 1855, chemistry professor Gottfried Wilhelm Ossan noticed during the electrolysis process the deposition of an acidic solute on the cathode of the apparatus. In similarity to the ozone formation that Schönbein had observed, he suggested that a second substance, this time containing hydrogen, was produced and released during electrolysis; he named it ozohydrogen (Ozone-Wasserstoff), but was never able to characterize it.

Schönbein's introduction of the concept of antozone led George Meissner to study another substance that did not exist and to draw several incorrect conclusions. What Meissner failed to do was to carefully analyze the chemical composition of the mist he considered to be antozone, which was formed during his laboratory experiments. It was not until the year 1870 that Engler and Nasse definitively showed that this mist was, in reality, hydrogen peroxide diluted in the aqueous phase.

Finally, at the beginning of the twentieth century, Carl Dietrich Harries, professor at Kiel who made important contributions on the chemical reactions of organic compounds with ozone, observed in the laboratory that the reaction between an alkane and an ozone molecule produced an organic substance that contained four oxygen atoms (instead of the three oxygen atoms provided by ozone). His hypothesis was that this unexpected product had to be formed from a yet unknown allotropic form of oxygen and that Harries called it oxozone. He suggested that this substance was the O_4 molecule.

Rubin highlights in one of his papers that none of these chemicals were isolated or characterized. Hence, they remained a myth for several decades and had many attributes of pathological science. Of course, as Schönbein wrote in 1858 in an article on antozone: "We philosophers cannot do any

investigation, chemist Carl Engler[15] (1842–1925) and his colleague physician
Otto Johann Nasse (1839–1903) brought the final proof in 1870: the antozone
observed by Meissner was not a modified form of oxygen, but rather, as
suspected, hydrogen peroxide.[16] In 1879, Albert Ripley Leeds (1843–1902),
professor of Chemistry at Stevens Institute of Technology, Hoboken, NJ, and
Vice President American Chemical Society, stated in his book, "The History
of Antozone and Hydrogen," that

> By far, the most important fact in the long and embarrassing history of
> antozone is the recent discovery that there is no antozone.

The Chemical and Optical Properties of Ozone

One of the soon recognized properties of ozone was its ability to disinfect
water. The rate at which this molecule could be produced in the laboratory
was, however, limited. A major technological breakthrough, however, was

---

15. Carl Oswald Viktor Engler became full professor of chemistry at the
Technische Hochschule of Karlsruhe in 1876 after having conducted research at the
universities of Freiburg and Halle. He specialized in the chemistry of mineral oil
(petroleum), and contributed to the development of the chemical industry in his
country. He was interested in the role of ozone as a disinfection agent for water. He
became a member of the Parliament (Reichstag) in 1887 where he represented the
national-liberal party.

16. "Zweifellos ist es alledem jedenfalls, dass das von Meissner aus dem
electrisirten Sauerstoff erhaltene und Atmizon oder Antozone genannte Gas nicht
eine besondere Sauerstoffmodification, sondern nur Wasserstoffsuperoxyd ist."
(In any case, it is undoubtedly true that the gas obtained by Meissner from the
electrified oxygen and called the atmozone or the antozone is not a special oxygen
modification, but only hydrogen peroxide.)

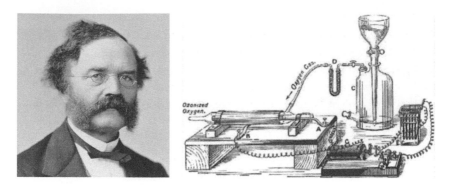

Figure 1.6. The device (right, reproduced from Fox, 1873) built and marketed by Werner von Siemens (left, reproduced from Corporate Knights Magazine, January 2017) to produce ozone. This invention paved the way for possible industrial applications of ozone including water disinfection.

achieved in Berlin as early as in 1857. German industrialist Ernst Werner von Siemens[17] (1816–1892) (Figure 1.6) and engineers in other countries developed and marketed the first ozone generator, which produced ozone levels at approximately 5% of the ambient oxygen content. Ozone was produced by generating electric spark discharges inside a tube through which oxygen was flowing. In Karlsruhe, Engler produced ozone in large quantities, also for water disinfection.

Ozone rapidly became a health-care agent, and a number of pilot projects were initiated to investigate the disinfection mechanism of ozone. In 1893, the first industrial scale ozone production facility was installed in Oudshoorn in the Netherlands. In the following years, chemist and engineer Marius-Paul Otto (1870–1939), who had completed in 1897 his doctoral thesis on the reactions of ozone with organic species at the Sorbonne, created in 1901 the "Compagnie Générale de l'Ozone" in Paris (Figure 1.7a). He designed an ozone-generating equipment, which was used around 1907 for the ozonation of the drinking water supply in Nice, his birthplace in southern France. Together with the "Société des Eaux de Bretagne," Otto's company became the "Compagnie des Eaux et de l'Ozone" specialized in water disinfection (Figure 1.7b). Different commercial companies were selling devices to purify water and air following the method established by Otto (Figure 1.8). In the United States, the first portable ozone generator was developed and patented in 1896 by Serbian-American Nikola Tesla (1856–1943), and his

---

17. Initiator of the electrical and telecommunications company Siemens.

Figure 1.7. (Top panel) Document showing the creation of Otto's "Compagnie de l'Ozone" in Paris in June 1901. Reproduced from https://numistoria.com/fr/services-des-eaux/14291-cie-de-l-ozone-brevets-et-procedes-m-p-otto.html. (Bottom panel) The building that hosted the "Compagnie des Eaux et de l'Ozone" in Paris at the beginning of the twentieth century. A company still operates today under the same name with its headquarter located at 21, rue de la Boétie in the 8th district of Paris.

Figure 1.8. (Top) Commercial published in 1907 by a company in Paris to promote water and air purifiers. The "ozoner" is sold to sterilize air and to treat pertussis (whooping cough), tuberculosis, anemia, and diabetes. (Bottom) Poster produced around 1930 to promote the purification of water by ozone (electrified air) using the method established by Otto. Reproduced from https://wellcomecollection.org/oembed/works/dc7488nu.

ozone devices aimed at purifying indoor air were distributed by the Tesla Ozone Company established in 1900. As a result, a commercial drinking water plant with ozone treatment was established in Niagara Falls in 1903. In Russia, the first water treatment plan was established in Saint-Petersburg in 1905. In 1909, ozone was adopted in Germany as food preservative for the cold storage of meat.

The decades following the discovery of Schönbein were devoted to the study of the chemical, thermodynamic, and optical properties of ozone. Ozone had become the molecule in vogue among chemists and many of them were trying to identify and study the reactions between this substance and various inorganic compounds, especially metals. The famous chemist von Liebig recognized the importance of Schönbein's work and wrote with a touch of arrogance:

> I consider the phenomena and observations described by this distinguished researcher to be as important as they are significant for science.[18]

Possible reactions of ozone with organic compounds were also considered, and Schönbein, himself showed as early as 1851 that ozone discolors indigo. In Kiel, Germany, Werner von Siemens' brother-in-law, Carl Friedrich Harris (1866–1923), undertook a systematic study of ozone reactions with organic compounds and introduced the term "*ozonide*" to characterize the products of these reactions.

In 1881, the first information on the optical properties of ozone appeared: the English chemist Sir Walter Noel Hartley (1845–1913) observed in his laboratory the very high absorption of ozone in the ultraviolet, more particularly between the wavelengths of 230 and 300 nm.[19] This absorption band, which today bears Hartley's name and was fully characterized by subsequent laboratory measurements, allows ultraviolet solar radiation that is harmful to living organisms to be completely absorbed before reaching the earth's surface.

---

18. "Ich betrachte die Erscheinungen und Beobachtungen, welche diese ausgezeichnete Forscher beschreibt, für ebenso wichtig wie bedeutungsvoll für die Wissenschaft."

19. One nanometer (abbreviated as nm) corresponds to 10^{-9} m.

CHAPTER TWO

The Presence of Ozone in the Atmosphere

Early Measurements of Ozone in Air

Schönbein's work had a great impact throughout Europe and was closely followed by the Académie des Sciences in Paris. In 1850, Antoine César Becquerel (1788–1878), professor of physics at the "Musée d'Histoire Naturelle" in Paris, member of the Academy and grandfather of the Nobel Prize winner Henri Becquerel (1852–1908) who discovered radioactivity, visited Schönbein in his laboratory in Basel. On his return to Paris, he presented to the Academy a paper "on the experiments of M. Schönbein on ozone."[1] Together with his colleague from the Ecole Polytechnique, the Academician Edmond Frémy, who was also interested in this molecule, he showed in 1852 that, if produced in a chamber whose open end was immersed in a potassium iodide solution (which the chemists note KI), ozone gradually disappears. The two researchers conclude that potassium iodide reacts with ozone (O_3), which produces molecular iodine (I_2) and potassium hydroxide (KOH).[2]

---

1. Communication relative aux expériences de M. Schönbein sur l'ozone.

2. $O_3 + 2KI + H_2O \rightarrow I_2 + 2KOH + O_2$ followed by I_2 + starch $\rightarrow$ violet complex.

The study of this particular reaction was important because it paved the way for the development of analytical methods by which the ozone concentration could be measured in the atmosphere. Schönbein developed an instrument that he marketed and made initial measurements by exposing for 12 to 24 hours to the air a sheet of paper impregnated with starch and potassium iodide (KI). A more or less intense bluish or purpled color appeared following the formation of a starched complex containing the iodine molecule (I_2) and resulting from the reaction of potassium iodide with ozone. The amount of ozone was then determined by comparison with a color scale provided with the instrument. The corresponding concentrations were assigned a value ranging from 0 to 10, and constituted what is known as the "*Schönbein scale*" (Box 2.1). The test paper, sometimes

---

Box 2.1 Units of the atmospheric concentration of ozone and other chemical species

The first determinations of the atmospheric abundance of ozone were based on the Schönbein method and the measured concentrations of ozone in air were expressed by a number on a scale ranging from 0 to 10, and based on the color of a paper impregnated with potassium iodide. Today, the ozone concentration is usually expressed as the number of molecules included in a cubic centimeter (or a cubic meter) of air. Typical values for the ozone *density* in the stratosphere are $1-5 \times 10^{12}$ molecules per cubic centimeter (molecules cm^{-3}). The atmospheric abundance of ozone can also be expressed as the mass of ozone per unit volume, thus, for example, in grams per cubic centimeter (g/cm^3) or kg per cubic meter (kg/m^3). In some case, it is expressed as the partial pressure of ozone and labeled in Pascal (Pa) or more often in milli-Pascals (mPa).

In many applications, the concentration of ozone is expressed relative to the total air concentration. The *volume mixing ratio* is defined as the ratio between ozone and air number densities. Similarly, the *mass mixing ratio* is defined as the ratio between ozone and air mass densities. The values are often expressed in parts per million (*ppm*, corresponding to a mixing ratio of 10^{-6}) or in parts per billion (*ppb*, corresponding to a mixing ratio of 10^{-9}). A suffix "v" or "m" is added to distinguish between the volume and mass mixing ratios. For example, the ozone mixing ratio is of the order of 1 to 8 ppmv in the stratosphere and 10 to 60 ppbv in the troposphere.

The ozone amount vertically integrated from the surface to the top of the atmosphere is expressed by the number of ozone molecules included in a vertical column of air whose horizontal base area is equal to 1 cm². Very often, the *ozone column* is expressed as the thickness of a pure ozone layer referred to standard pressure and temperature. This thickness is typically of the order of 2.5 to 5 mm, dependent on the latitude and season. A specific unit for the ozone column has been introduced by British scientist Gordon Dobson: a column of 3 mm, for example, corresponds to 300 *Dobson units* (or DU).

The definitions provided here for ozone apply to any other chemical species in the atmosphere.

called an ozonoscope, was usually placed inside an enclosure that protected it from the direct effects of sunlight (Figure 2.1). However, the method was not very accurate because it was strongly influenced by the humidity of the ambient air, the type of paper adopted, the exposure time of the paper, the wind speed, and by the action of other chemical compounds present in the atmosphere: the measured concentrations were overestimated by the presence of nitrogen dioxide (NO_2) and hydrogen peroxide (H_2O_2) in the air and underestimated by the presence of sulfur dioxide (SO_2) and ammonia (NH_3). The method was therefore more qualitative than quantitative. A large number of alternative methods were therefore proposed. In some of them,

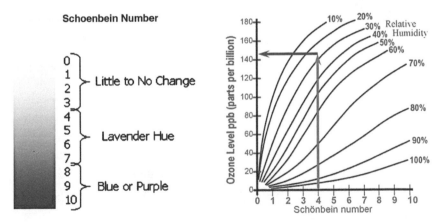

Figure 2.1. (Left) Schönbein color scale used to quantify measured ozone concentrations in the air. (Right) Conversion of the Schönbein number into an ozone concentration expressed in ppb (parts per billion) for different values of the relative humidity. Source: Environmental Protection Agency, https://www3.epa.gov/airnow/flag/Field-Testing-for-Ozone-LP.pdf.

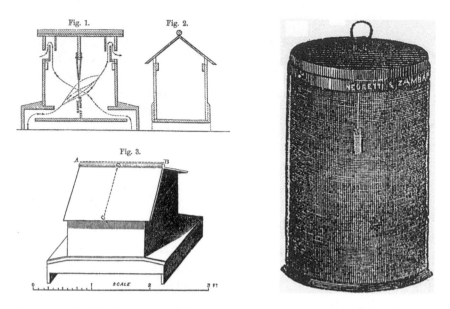

Figure 2.2. (Left) Enclosure imagined by W. F. Moffat, which exposes the test paper to airflow without any possible disturbance by solar radiation. The interior of the enclosure is painted black. The ozone sensitive paper is located in the middle of the enclosure. Moffat measured the ozone content, particularly at sea, around 1867. (Right) The "cage" built by Sir James Clarke, which consists of two cylinders of fine gauze threads. The test paper was suspended in the center of the inner cylinder. Reproduced from Fox (1873).

the proportion of potassium iodide relative to starch was modified to make the measurement more sensitive to the atmospheric ozone content.[3]

The choice of the method was not unanimously accepted so that an English meteorologist described the measurement as "a delusion and a snare." Other scientists pointed out that the color of the test paper varied with time during the measurement; it turned blue and then turned white again for no apparent reason. The results also varied substantially according to the measurement enclosure used (Figure 2.2) and the operator in charge of the measurement. This led the Scottish Meteorological Society to state in 1869 that there is no method to conclude with certainty that ozone and ozone alone is the coloring agent in the test paper. Already four years earlier, in 1865,

---

3. Different proportions were suggested by different experimentalists. The ratio of potassium iodide content to starch content was 1/10 in the method proposed by Schönbein, 1/2.5 in the method proposed by Moffat, and 1/5 in the method suggested by Lowe.

the Académie des Sciences de Paris, also unconvinced of the quality of the measurements, established a nine-member committee to limit the number of articles on ozone measurements submitted to the prestigious institution, and to make recommendations for improving the quality of the scientific papers to be published by the French Academy. Among the members of this committee was the famous astronomer Urbain Le Verrier (1811–1877), who had discovered the existence and position of planet Neptune and whom the popular author of astronomical books, Camille Flammarion (1842–1925), described as a haughty, disdainful, and intractable autocrat.

In 1858 (almost 20 years after Schönbein's detection of the ozone's odor), an important discovery was made by French agronomist and chemist, Jean Auguste Houzeau (1829–1911; Figure 2.3). He, professor at the "Ecole Supérieure des Sciences et des Lettres" in Rouen and at the "Ecole d'Agriculture du Département de la Seine Inférieure," detected the presence of "odorous oxygen" in the air by observing the reaction of this gas with a mixture of arsenic and iodine, and concluded that ozone is a permanent chemical component of the atmosphere.

Figure 2.3. Jean Auguste Houzeau who showed that ozone is a permanent chemical constituent of the atmosphere. Reproduced from https://seine76.fr/celebrites76/popup_portraits.php?var_portrait=houzeau-auguste.

This discovery must have delighted Schönbein, who had sensed the presence of ozone in the atmosphere several years earlier, at least during the occurrence of thunderstorms. In fact, Schönbein had detected several times the odor of ozone in the air, in particular on the day during which lightning had struck the bell tower of a church near his home in Basel.[4] However, the source of atmospheric ozone remained a disputed issue during the second half of the nineteenth century. Based on his own observations, Schönbein was convinced that atmospheric ozone is formed primarily by lightning strikes, but this hypothesis was strongly disputed by Houzeau, who stated that the lightning strikes are more likely to produce nitrous acid and nitric acid than ozone. This last statement by Houzeau is interesting since we know today that lightning flashes produce substantial quantities of nitric oxide and that this chemical species contributes to the formation of nitric acid and ozone in the troposphere.

Systematic Surface Ozone Measurements

With the development of the measurement techniques described above, several hundred ozone observing stations were established throughout Europe and in the United States, particularly between 1860 and 1900. At that time, as will be mentioned later in this chapter, people were concerned by the possible influence of ozone on human health. Hence, it was important to monitor the abundance of this gas in the atmosphere. Even in a small country like Belgium, more than 150 stations were established and provided data for more than a decade. In Paris, several stations provided ozone measurements and highlighted the differences in the concentrations observed in the different districts of the city (Figure 2.4).

The analysis of the data provided by the ozonoscopes revealed the existence of daily and seasonal variations in the ozone content at the Earth's surface. Houzeau recorded for a period of ten years (1861–1870) the number of days during which the ozone content was high, which allowed him to draw some interesting conclusions. On average in Rouen, the number of days with high ozone values was 12.6 during the three winter months, 40.5 during the spring, 37.4 in the summer, and 19.3 in the autumn. He could therefore deduce that ozone is most abundant in spring, and that the surface concentration of this gas is minimum in November. The measurements made near Emden

---

4. Around 900 BC, Greek poet Homer had already reported a peculiar smell after lightning storms.

Stations.	Mean of Daily Observations during the years 1866 and 1867.
Passy	(a) 6·39
Monceau	4·04
Montmartre . . .	4·48
La Villette . . .	(b) ·96
Charonne	4·34
Ménilmontant . . .	(c) 1·16
Boulevard de Picpus . .	4·49
La Boule-Rouge . . .	2·45
Fontaine-Molière . . .	(d) ·38
École de Médicine . .	(e) ·80
Rue Racine . . .	(f) 1·69
Panthéon	2·83
Saint-Victor . . .	4·98
Boulevard d'Italie . .	3·08
Vaugirard	·89
Réservoir de Vaugirard . .	(g) 8·37
La Chapelle . . .	(h) 3·08
Butte-aux-Cailles . . .	(h) 4·79
	1866.
Batignolles . . .	4·75

(a) Near resinous trees.
(b) Close to a quay on the Seine.
(c) Near a tallow manufactory.
(d) A public urinal is situated almost directly underneath the test.

[1] *Bulletin de Statisque Municipale*, February 1868.

Figure 2.4. Ozone measurements in several districts of Paris in 1866 and 1867. The values that vary between 0.38 and 8.37 correspond to a color (Schönbein) scale used at the time. The low values measured at Fontaine Molière are probably due to the presence of high levels of ammonia in the air at this particular location, which introduces an error in the ozone measurement. The data are taken from the Bulletin de la Statistique Municipale de Paris published in February 1868. Reproduced from Fox (1873).

in northern Germany (Figure 2.5) highlighted the differences between day and night concentrations, and indicated the existence of a similar seasonal variation in the ozone abundance near the ground: ozone concentrations were highest in March (with a second peak in September) and lowest in July and November. The cause of these variations was not understood.

As already mentioned, measurements made with paper impregnated with potassium iodide were unreliable because they were affected by the presence of reactive gases other than ozone in the atmosphere. Thus, they did not provide any reliable quantitative information on the concentrations of surface ozone. At most, they provided some qualitative indications and therefore do not constitute a credible historical reference.

The most accurate historical data that sometimes serve as a reference to describe the ozone content at the earth's surface during the preindustrial era are provided by the measurements made between 1876 and 1910 at the

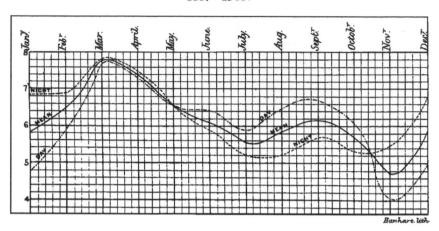

Graphical Delineation
of the difference between and the amount of Day and Night
Ozonic Reaction during the various Months of the Years
1857 - 1863.

Figure 2.5. Change in the seasonal evolution of ozone content measured at Emden in Germany near the North Sea. Values averaged over the period 1857 to 1863 are expressed in terms of the Schönbein color scale. Based on these observations, the surface concentration was highest in March and lowest in November. The graph also shows the differences between day values (higher in summer) and night values (higher in winter). From Fox (1873).

Meteorological Observatory of the "Parc Montsouris" located at the southern edge of Paris (Figure 2.6). The Observatory was established in 1867 by Charles Joseph Sainte-Claire Deville (1814–1876) to monitor weather and air pollution in the French capital. In 1873, Albert Lévy (1844–1907; Figure 2.7), a former student of Ecole Polytechnique was hired by the Observatory to address questions related to public hygiene and sanitation in Paris. In 1887, he became the Head of the Chemistry Department of the Observatory, which gave him a unique opportunity to develop new instrumentation to measure different gases in the atmosphere including ozone. Albert Lévy is described as an inspiring, rigorous, generous, and independent personality, whose passion was to educate and support his students at the School of Physics and Chemistry established in Paris in 1882. Albert Lévy also wrote several popular science books on issues related to physics, astronomy, meteorology, and air chemistry.

The device developed by Albert Lévy to measure ozone was a chamber into which a known quantity of ambient air was bubbling through a 3% aqueous solution of potassium iodide (KI) in a vessel filled with an arsenic compound (K_3AsO_3). The ozone content in the air was derived from the

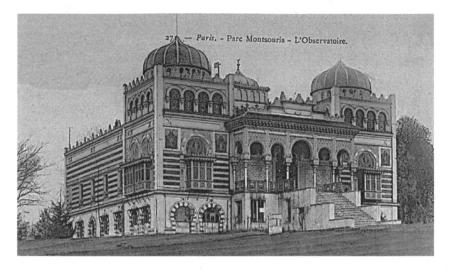

Figure 2.6. The building hosting the Observatory of the Parc Montsouris in Paris around 1900. This facility, called Palais du Bardo, was a replica of the Palace of the Bay of Tunis. It was originally built for the 1867 Paris Universal Exhibition, and was relocated in 1869 in the Parc Montsouris. The observatory, created in 1873 by geologist and meteorologist Charles Sainte-Claire-Deville (1814–1876), primarily to perform meteorological observations, moved in the building of Parc Montsouris in 1876. Albert Lévy joined the observatory on May 7, 1873 and became the Head of its Chemistry Department in 1887. The Department was responsible for the measurements of atmospheric ozone and for bacteriological analyses of drinking water in the French capital. The building made of wood and stucco lacked maintenance and was gradually damaged and eventually abandoned. It was completely destroyed by a fire in 1991. Source: Postcard reproduced from http://paris1900.lartnouveau.com/paris14/parc_montsouris/palais_du_bardo.htm.

measurement of the conversion rate by ozone of arsenite (AsO_3^{3-}) to arsenate (AsO_4^{3-}), with potassium iodide acting as a chemical catalyst. The concentrations derived by this method and systematically reported in the "Annales de l'Observatoire Municipal de Montsouris" were in the range of 5 to 15 parts per billion (ppbv), that is, at least a factor of three lower than the values currently recorded in Europe (Figure 2.8). A recent reevaluation of the laboratory method[5] shows that interference from non-ozone oxidizers was minor

---

5. In 1988, two German scientists Andreas Volz and Dieter Kley, both at the Helmholtz Research Center of Jülich, reproduced the instrument constructed by of Albert Lévy and showed that the measurements made by this system installed at the Montsouris Observatory were accurate. This study concluded that the average ozone concentration in Paris at the beginning of the twentieth century was close to 11 ppbv.

Figure 2.7. Albert Lévy, Head of the Chemistry Department of the Montsouris Observatory in Paris, performed systematic measurements of the surface ozone concentration in Paris around 1900. Source: Albert Lévy 1844–1907, Notice nécrologique, Annuaire de l'Association Amicale des Anciens Elèves du Lycée Charlemagne, Paris 1908.

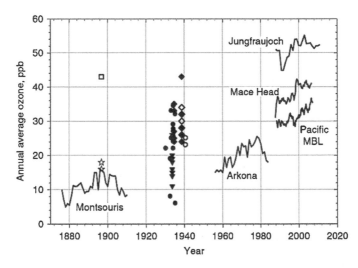

Figure 2.8. Annual average values of ozone concentration expressed in ppbv measured at Parc Montsouris in Paris between 1876 and 1910, and, in the decades that followed, at Arkona on the German island of Rütgen in the Baltic Sea. The symbols refer to individual measurements in the Alps (Grand Mulets [square], Chamonix [stars], Arosa [circles], Lauterbrunnen [triangles on the tip], Jungfraujoch [diamonds], and Pfänder [circles] using chemical [open symbols] or spectroscopic [full symbols] methods). The low values reported at Parc Montsouris at the end of the nineteenth century and at the beginning of the twentieth century are believed to be underestimated by perhaps a factor 2. Reproduced from Calvert et al. (2015).

in relatively clean air, but that the measurements were disturbed when the air was polluted. After carrying out measurements for several years, Albert Lévy himself realized the existence of an interference by the "gaz reducteurs" (reducing gases) present in the atmosphere, and a correction was therefore applied to take this effect into account.

In his book on the "History of Air" published in 1879, he wrote,[6]

There is generally more ozone in the air of the countryside than in that of the cities, and it has sometimes been explained by the greater healthiness of the countryside. One has noticed the absence of ozone in the air at the time of cholera epidemics; but these very interesting studies undoubtedly need to be reconsidered. The wind direction has a strong influence on the amount of ozone in the air. Thus, at Montsouris, we have seen this year (1877–1878), as we did last year, that every time the wind turns to the south the amount of ozone increases; each time, on the contrary, when the wind turns to the north, the quantity of ozone decreases. If we observe that the observatory is located south of Paris, that the north winds have passed over the city, while the south winds come from the countryside, we can think that the air of Paris

6. On trouve généralement plus d'ozone dans l'air des campagnes que dans celui des villes et on a quelquefois expliqué par cette raison la salubrité plus grande des campagnes. On a cru reconnaitre l'absence d'ozone dans l'air au moment des épidémies de choléra; mais ces études très intéressantes sans doute, ont besoin d'être reprises et suivies avec soin. La direction des vents exerce une influence marquée sur la quantité d'ozone contenue dans l'air. Ainsi à Montsouris, nous avons constaté cette année (1877–1878), comme nous l'avions fait l'an dernier, que chaque fois que le vent tourne au sud la quantité d'ozone augmente; chaque fois au contraire que le vent tourne au nord, la quantité d'ozone diminue. Si nous observons que l'observatoire est situé au sud de Paris, que les vents du nord ont passé sur la ville, tandis que les vents du sud viennent de la campagne, on pourra penser que l'air parisien explique ces différences. On remarque en outre un phénomène assez curieux: un papier ozonométrique déjà coloré en bleu par suite de l'action de l'ozone, se décolore lorsque le vent passe au nord, comme si l'air contenait en ce moment un principe spécial qui décompose l'iodure d'amidon formé en régénérant l'iodure de potassium [...]. Il n'est peut-être pas inutile, à ce propos, de rappeler que Théodore de Saussure a montré qu'il existe dans l'air une substance carbonée qui n'est pas l'acide carbonique. Ces expériences ont été confirmées par M. Boussingault, qui a trouvé qu'indépendamment d'un corps carburé, l'air contenait un corps hydrogéné. Plusieurs chimistes pensent que ce principe hydrogéné et carboné est peut-être le carbure d'hydrogène, connu sous le nom de gaz des marais dont il se dégage dans l'air des quantités considérables.

explains these differences. There is also a rather curious phenomenon: an ozonometric paper already colored blue by the action of ozone, fades when the wind passes to the north, as if the air contained at this moment a special principle which decomposes starch iodide to regenerate potassium iodide [...]. It may not be useless, in this connection, to recall that Theodore de Saussure[7] has shown that there exists in the air a carbonaceous substance that is not carbonic acid. These experiments were confirmed by M. Boussingault,[8] who found that, independently of a carbonated body, the air contained a hydrogenated body. Several chemists think that this hydrogenated and carbonated principle represents perhaps the carbide of hydrogen, known under the name of marsh gases whose considerable quantities are released in the air.

There is increasing evidence today that the values of approximately 10 ppbv quoted at the end of the nineteenth century represent an underestimation of the surface ozone concentration. At that time, Paris was a city of 2.5 million inhabitants and the energy supply came primarily from coal burning. The measurements made at the Montsouris Observatory, located at the southern edge of the city, were directly affected by the substantial quantities of sulfur dioxide (SO_2) produced by coal burning, particularly when the wind was blowing from the north. SO_2 has a negative influence on KI ozone measurements because it reduces iodine (I_2) to iodide KI. Gases produced by agricultural operations south of Paris including ammonia also generated measurement errors, primarily when the wind was blowing northwards. Large amounts of ammonia were also produced in the city by the 80,000 horses and 5,700 dairy cattle that were present in the French capital city at that time. A recent analysis of the early ozone observations made by an international group of experts and led by David Tarasick[9] from Environment and Climate Change Canada and Ian Galbally from the Commonwealth Scientific and Industrial Research Organisation (CSIRO) has shown that the preindustrial surface concentration of ozone was most likely closer to 20 ppbv than to 10 ppbv.

---

7. Nicolas Théodore de Saussure (1767–1845), Swiss chemist and plant physiologist.

8. Jean-Baptiste Boussingault (1801–1887), French chemist interested in agricultural and petroleum science.

9. Tarasick et al., Tropospheric ozone assessment report: Tropospheric ozone observations—How well can we measure tropospheric ozone today and in the historic past, *Elementa: Science of the Anthropocene*, 7 (2019), 39. DOI: http://doi.org/10.1525/elementa.376

Systematic measurements of ozone, initiated in 1871 by R. C. Kedzie, a scientist at the Agricultural College of the State of Michigan in Lansing (now Michigan State University), and performed as part of a statewide network of approximately 20 stations, provide a picture that is very different from what was reported in Europe. Observations made using the Schönbein test paper were characterized by strong daily variations in the surface ozone concentrations. When converted in ppbv using the Schönbein scale displayed in Figure 2.1 and accounting for the relative humidity inside the measurement chamber, the reported values ranged from about 10 to 100 ppbv (Figure 2.9). The highest ozone levels occurred when the wind was blowing from the southwest. Concentrations appeared to increase as the air masses progressed over the industrial and urbanized areas of the state of Michigan, but they dropped abruptly after the passage of a cold front. These patterns were not different from those observed in today's atmosphere.

Current mathematical models that simulate the chemical composition of the atmosphere under preindustrial conditions suggest that the levels were more likely close to 20 to 25 ppbv in the late 1800s and early 1900s, consistently with the recent assessments of early observations. Today, the surface ozone concentrations influenced by human activity are, for example, in the range of 30 to 45 ppbv at the Mace Head station on the west coast of Ireland (Figure 2.8). In central Europe, concentrations measured at the Jungfraujoch station in Switzerland, at an altitude of 3,454 m above sea level, reach average values that vary with the season between 40 and 60 ppbv. These values, however, are much lower than the peak ozone levels observed in some urban areas, including Los Angeles, where maximum concentrations of 600 ppb were recorded in the 1950s and 1960s, and today still reach values of 150 to 200 ppb, well above the limits imposed by US regulations.

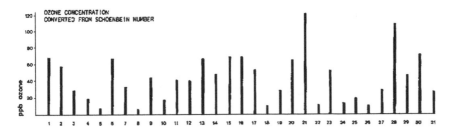

Figure 2.9. Daily measurements of the surface ozone concentration (ppbv) made at Michigan Agricultural College from July 1 to 31, 1879. The Schönbein paper was exposed to ambient air each day from 7 am to 2 pm. Reproduced from Linvill et al., *Monthly Weather Review* 108 (1980), 1883–1880.

Ozone and Health

Between 1860 and 1900, the prevailing idea was that the presence of ozone in the air could prevent cholera epidemics. For example, in the French city of Lyon, the so-called "ozone free" city, cholera cases were found to be more frequent than elsewhere in the country, and the disease more intense than in other French cities. The perception at that time of the relationship between ozone and health is well described in the book by Dr. Cornelius Fox, a member of the Royal College of Physicians in London and published in 1873 under the title *"Ozone and Antozone. Their History and Nature; When, where, why and how is ozone observed in the atmosphere?"* (Figure 2.10). In his book, Fox, who had discovered that ozone destroys microorganisms, claimed that ozone is a "purifying agent of the highest order" and stated that this gas

> should be pumped into our mines and cities, and diffused through fewer wards, sick rooms, the crowded localities of the poor or wherever the active power of the air is reduced and poisons are generated. Its employment is especially demanded in our hospitals, situated as they mostly are in densely populated districts, where the atmosphere is nearly always polluted by rebreathed air, decomposing substances and their products, and where no mere ventilation can be fully effective. If practicable, it would be highly advantageous to direct streams of sea air, or air artificially ozonized into the fever and cholera nests of our towns. Ozone may be easily disseminated through public buildings, theatres and other confined atmospheres, where numbers of people are accustomed to assemble, in order to maintain the purity of the air.

Such claims could not be made today. The book also contains specific references to medical studies that point out, for example, that malaria and, in general, the appearance of fevers was favored by the absence of ozone in the air. Some reports cited in this book claimed that an increase in human mortality was observed when the atmospheric ozone content decreases, while other reports did not link the presence of ozone to the number of lung diseases. In a book published in the German city of Halle in 1879 under the title *"Historische-kritische Studien über das Ozon"* (Historical-critical studies on ozone), the chemist from Karlsruhe C. Engler provides a good summary of the knowledge at this time.

The idea that ozone can be used as a remedy for various diseases gained therefore ground, even though no convincing scientific explanation had been provided. For example, in 1880 a sanatorium established by Dr. John H. Kellogg in Battle Creek, Michigan, began offering medical cures against

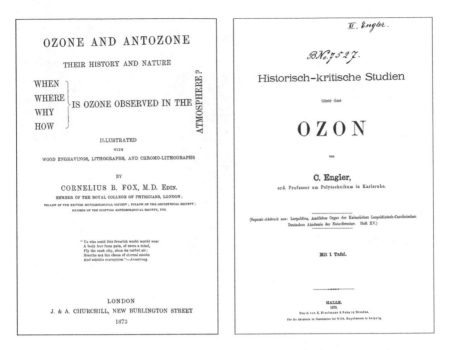

Figure 2.10. (Left) Book published in London (1873) by Scottish physician Cornelius Benjamin Fox on "Ozone and Antozone." (Right) Book published in Halle, Germany (1879) by Carl Engler, professor at the Polytechnikum of Karlsruhe that reviews "historical-critical studies about ozone."

diphtheria, using saunas with ozonized steam. Later, in 1898, an institute for oxygen therapy was established in Berlin by two physicians, Dr. Thauerkauf and Dr. Luth. A real craze was developing, and getting treated by ozone became fashionable in bourgeois society. In 1926, Otto Heinrich Warburg (1883–1970), the Director of the Kaiser Wilhelm Institute for Cell Physiology in Berlin-Dahlem, suggested that the cause of cancer was the lack of oxygen in human cells, which led people to believe that the injection of ozone in the human body would be a cure against this disease. For his work, Warburg (Figure 2.11) was awarded the Nobel Prize for Medicine in 1933. The perceived importance of his discovery was such that he was authorized by the National Socialist regime to continue his research despite his Jewish origin,[10]

10. One reason was that the mother's family of Warburg was not Jewish and another was that Reichs-Marshall Göring, who had the right to decide who was a Jew, ruled that he was a quarter-Jew. As a result, Warburg, who could continue his research, was not allowed to teach in a university or to hold any administrative

Figure 2.11. Nobel Prize laureate Otto Heinrich Warburg studied chemistry in Freiburg and Berlin, and later medicine in Berlin, Munich, and Heidelberg, where he completed his doctorate in 1911. He was associate professor of physiology in Berlin from 1921 to 1923, and founded the Kaiser Wilhelm Institute for Cell Physiology in 1930. Source: Bundesarchiv_ Bild_102-12525, Otto_Heinrich_Warburg.

and was awarded a second Nobel Prize in 1944. He remained Director of his Institute until 1967. Warburg wrote,

> Cancer, more than other diseases, has multiple causes. In a nutshell, the main cause of cancer is the replacement of oxygen respiration in normal body cells with sugar fermentation."

Today, some groups continue to promote ozone therapy to cure cancer, AIDS, multiple sclerosis, Alzheimer's and Lyme diseases as well as dental infections, and even to repair herniated discs. In the past, ozone, when used as a therapy, was administered into the human body in gaseous form by buccal, rectal, or vaginal injection, or in liquid form by intramuscular or intravenous shots (Figure 2.12). Sometimes blood taken from patients and exposed to ozone was reinjected into the blood vessels. The effectiveness

---

position. He was denounced on several occasions for having criticized the Nazi regime, but was somewhat protected probably in the hope that he would find a cure to cancer.

Figure 2.12. (Left) Ozone therapy provided in the early 1900s by Armenian-American radiologist Mihran Krikor Kassabian who stated that the therapy is of "paramount value where sprays or medicated vapors cannot reach the part by other means." (Right) Pin distributed around 1900 to promote *Liquozone*, formerly "Powley's Liquified Ozone" as a defense against germ-related illnesses. Since liquid ozone forms only at temperatures below 112°C, the product was clearly a fraud. From Natalie Jacewicz, Science History Institute and https://www.sciencehistory.org/distillations/a-killer-of-a-cure

of these treatments has never been scientifically proven, and the treatment is therefore often presented as an "alternative" therapy. Schönbein, himself noted that inhaling ozone could cause chest pains and breathing difficulty, and Scottish chemist James Dewar (1842–1923) noted in 1874 that frogs and birds exposed to ozonized air experienced severe respiratory problems.

As early as 1856, 17 years after its discovery, ozone was used as a disinfectant of operating rooms, and in 1892 the British medical newspaper *The Lancet* suggested adopting ozone as a cure for tuberculosis. During the First World War, ozone was used in Queen Alexandra's Military Hospital in London to disinfect the wounds of soldiers returning from the front, and the US Army Medical Department also included ozone as a key method for reducing war-related infections. Even today, several associations are calling for the practice of what they call Warburg's method to cure cancer. They claim that some governments influenced by pharmaceutical companies refuse to recognize the benefits of this simple and inexpensive therapy. This statement, however, is not shared by many specialists in the medical profession who consider, to the contrary, that "ozone therapy" can be dangerous for the human body and can endanger the lives of patients. For this reason, the US Food and Drug Administration (FDA) banned the use of ozone for therapeutic purposes in 2016, noting that "ozone is a toxic gas with no known useful medical application in specific, adjunctive, or preventive therapy. In order to be effective as a germicide, it must be present in concentrations far greater than that which can be safely tolerated by man and animals." And as stated by freelance science writer Natalie Jacewicz[11]:

> For more than a century ozone therapy has been a source of false hope for the sick and ill-gotten gains for the crooked.

Attempts to Detect Antozone in the Atmosphere

Around 1860, while Schönbein was trying to convince the chemists' community of the existence of antozone, several experimenters attempted to detect the presence of this substance in the atmosphere. In 1860, A. Mitchell in the United Kingdom suggested that the discoloration of the measurement papers in ozonoscopes, which was occurring around 6 or 7 o'clock in the morning, could be due to the presence of antozone in the atmosphere. He even indicated that this observation provided a proof of the existence of this substance. The problem was widely discussed because several scientists believed that the observed discoloration of the ozonoscope paper should be attributed to a chemical compound different from ozone that is also present in the atmosphere. Some even pointed to the possible presence of hydrogen that could disturb the measurement of the ozonoscope. The antagonist substance that perturbed the measurement of ozone received the name of "true antozone" to distinguish it from Schönbein's antozone, but without ever establishing its chemical nature. It was believed at this time that some oxygen could exist in a state opposite to that of the ozonized state, but nothing proved that this unidentified compound would discolor the ozonoscopes' papers. Some fanciful observations suggested that the "true antozone" was most abundant when the wind was blowing from the north. And since antozone was considered to be an "electrified" substance, the prominent Belgian astronomer and mathematician Adolphe Quételet (1796–1874), who was the perpetual secretary of the Belgian Royal Academy for 40 years, linked this assertion to the "quality of atmospheric electricity."

---

11. https://www.sciencehistory.org/distillations/magazine/a-killer-of-a-cure.

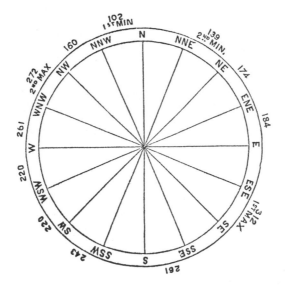

Figure 2.13. Diagram drawn up by Belgian astronomer Adolphe Quételet, which shows that the dominant winds of the North (N to NNW), rich in antozone, correspond to the first minimum of the "electric force." Reproduced from Fox (1873).

He indicated that, on average, the "electric force" in the atmosphere is minimal when the prevailing winds originate from the north and north-northwest directions or from a point situated between the geographic and magnetic poles (Figure 2.13). This corresponded to a situation in which the concentration of antozone is high. All these considerations no longer applied once the concept of antozone was definitely abandoned in 1879.

CHAPTER THREE

Spectroscopic Determinations of Ozone

Spectroscopic Properties of Ozone

An important discovery regarding the optical properties of atmospheric ozone was made in 1879 by Marie Alfred Cornu (1841–1902; Figure 3.1), a professor at the Ecole Polytechnique de Paris who observed that no solar radiation at a wavelength shorter than 293 nm reaches the Earth's surface. Cornu could not explain this observation. The measurement, performed two years later (1881) by Sir Walter Hartley (1845–1913) (Figure 3.1) in his laboratory at the Royal College of Science in Dublin where he had been appointed professor in 1879, provided a plausible explanation. The British chemist showed that ozone absorbs intense ultraviolet radiation at wavelengths shorter than 285 nm. He wrote in 1881 about his photographic measurements of the ozone absorption spectrum:

> The photograph taken showed a broad absorption band stretching from wavelength of about 285 to 233 millionths mm [...] The mean wavelength intercepted by ozone is 256 millionths mm.

Hartley explained Cornu's observation, and in particular the existence of a very clear spectral limit at 293 nm, by the presence of ozone in the atmosphere, probably at high altitude. And, in an article published the same

Figure 3.1. (From top to bottom and from left to right) Marie Alfred Cornu, (credit: Meisterdrucke; https://www.meisterdrucke.de/kunstdrucke/Nadar/295243/ Marie-Alfred-Cornu-(18411902).html), Sir Walter Hartley (credit University College Dublin; http://www.ucd.ie/merrionstreet/1910_hartley.html), Sir William Huggins (credit: Wikisource; https://en.wikisource.org/wiki/Author:William_Huggins), Lady Margaret Huggins (Reproduced from Barbara J. Becker, *Astronomy & Geophysics*, 57 (2016), 2.13–2.14), and Oliver Reynolds Wulf (credit: California Institute of Technology, Pasadena, CA) studied the spectroscopic properties of ozone.

year, he stated that ozone is a normal constituent of the upper atmosphere, in greater proportion at high altitude than near the Earth's surface. And he added that

> the amount of ozone in the atmosphere must be sufficient to explain the limitation of the solar spectrum in ultraviolet without considering the possible absorption by major atmospheric compounds, oxygen and nitrogen.

Further work conducted in the laboratory soon provided additional information. The French chemist Louis Philibert Claude James Chappuis

(1854–1934), known as James Chappuis, a graduate of the Ecole Normale Supérieure and after 1881 a professor at the Ecole Centrale des Arts et Manufactures in Paris, noted that, when white light passes through an enclosure containing ozone, the gas takes on a bluish color. Chappuis deduced that ozone must absorb some of the visible radiation, mainly the red, orange, and yellow colors. Chappuis, although not totally convinced, imagined that the blue color of the sky could be due to the presence of ozone. Such a hypothesis was restated more convincingly a year later by Hartley. We know today, thanks to the work carried out in 1913 by John William Strutt (third Baron Rayleigh, 1842–1919), that the blue color of the sky is in fact the result of the scattering of sunlight by the molecules present in the atmosphere. As a matter of facts, Chappuis' studies allow us to quantify the absorption of radiation in the visible part of the spectrum. In a note published in 1880, Chappuis wrote,

> The spectrum of ozonized oxygen by the electric effluent observed with a single- or two-prism spectroscope shows eleven distinct bands of absorption in the usual visible part of the spectrum. I drew up the map of these bands and compared it to the maps of the telluric bands.[1]

Further progress was being made in the following years. The English astronomer Sir William Huggins (Figure 3.1), who had built his observatory on his private property in London (Tulse Hill), observed spectral lines emitted by various celestial objects. Together with his wife Lady Margaret Lindsay Huggins, who contributed to his observations, he identified in 1890 several groups of absorption lines in the Sirius and Vega spectra between the wavelengths of 320 and 334 nm. The origin of these lines was not known, but Huggins related them to the presence of an absorber in the Earth's atmosphere. The lines were eventually attributed to ozone in 1917 by astronomer Alfred Fowler (1868–1940) and physicist Robert John Strutt (fourth Baron Rayleigh; 1875–1947), who worked together at Imperial College in London. An accurate measurement of the absorption coefficients made by two French spectroscopists, Maurice Paul Auguste Charles Fabry (1867–1945), known as Charles Fabry, and Henri Buisson (1873–1944), in

---

1. "Le spectre de l'oxygène ozonisé par l'effluve électrique observé à l'aide d'un spectroscope à un ou deux prismes présente onze bandes d'absorption bien nettes dans la partie ordinairement visible du spectre. J'ai dressé la carte de ces bandes et je l'ai comparée aux cartes des bandes telluriques."

their Marseilles laboratory had already indicated in 1913 that there was a coincidence between the spectrum of ozone absorption and the atmospheric attenuation of solar and stellar radiation. By observing the solar spectrum between 300 and 334 nm at varying solar zenith angles, they were able to prove that the absorbing substance affecting these wavelengths was present in the earth's atmosphere. This allowed the two French researchers to definitely establish the presence of ozone at high altitudes, and so to confirm the hypothesis formulated 32 years earlier by Hartley.

Finally, the American chemist and meteorologist Oliver Reynolds Wulf (1897–1987), who was interested in the structure of the ozone molecule, first at the California Institute of Technology in Pasadena, California, and then at the Bureau of Chemistry and Soils of the US Department of Agriculture in Washington, discovered in the 1920s a series of ozone absorption bands in the visible and near-infrared spectral regions. The ozone absorption spectrum (Figure 3.2) established by the pioneers of ultraviolet spectroscopy shows that radiation is strongly absorbed at wavelengths shorter than 300 nm. The spectral information obtained from laboratory measurements provided the basis for the measurement of ozone in the atmosphere.

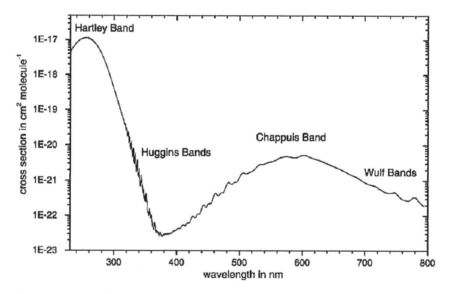

Figure 3.2. Ozone absorption spectrum with the Hartley band and the Huggins bands in ultraviolet, the Chappuis band in the visible, and the Wulf bands in the near infrared. Reproduced from Orphal, J., *Journal of Photochemistry and Photobiology, A Chemistry* 157 (2003), 185–209.

The Discovery of Ozone in the Stratosphere

At the beginning of the twentieth century, Charles Fabry (Figure 3.3) played a major role in the research on the ozone layer. After successfully completing his secondary studies at the Lycée de Marseilles, he was admitted to the Ecole Polytechnique de Paris in 1885 at the age of 18. He then returned to Marseilles where he was accepted in 1889 for the competitive "aggregation" examination. He first devoted himself to teaching in secondary schools and was appointed professor of physics in various high schools, successively in Pau, Nevers, Bordeaux, Marseilles, and Paris. At the same time, he prepared a doctoral thesis, which he defended at the University of Paris in 1892 under the title "Théorie de la visibilité et de l'orientation des franges d'interférences" (Theory of the Visibility and Orientation of Interference Fringes). His theoretical work, which immediately gave him authority among specialists in optics, spectroscopy, and photometry, served as the basis for the development of a new measurement method.

Fabry joined the University of Marseilles in 1894 where he first became senior lecturer and later in 1904 was awarded the title of professor. During this period, he worked with his colleague, also a former student at Ecole Polytechnique, Jean-Baptiste Gaspard Gustave Alfred Perot (1863–1925; called Alfred Perot), and together they developed the interferometer which bears their name (Fabry-Perot) and which is still used today for many scientific and technical applications. Both men worked well together, and Fabry later wrote about his friend who had just died:

> I still see Perot at the beginning of his scientific career with his tireless activity, his open-mindedness, his exceptional ability to work by hand, constructing with his hands the devices necessary for his research, communicating his sacred fire to those around him.[2]

Fabry's studies of atmospheric ozone were conducted in cooperation with Henri Buisson (Figure 3.3), former student of the Ecole Normale Supérieure, and full professor at the University of Marseilles from 1914. Both scientists were investigating the absorption of ozone in the ultraviolet with the purpose of explaining why solar radiation at wavelength shorter than about 300 nm

---

2. "Je revois encore Perot au début de sa carrière scientifique avec son inlassable activité, son esprit ouvert, son exceptionnelle habileté au travail manuel, construisant de ses mains les appareils nécessaires à ses recherches, communiquant son feu sacré à ceux qui l'entouraient."

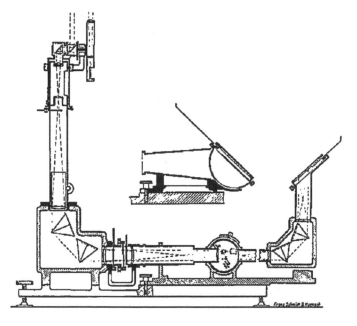

Figure 3.3. Maurice Paul Auguste Charles Fabry (upper panel) and the spectrometer (lower panel) that, together with Henri Buisson, he used to make the first measurement of the ozone layer thickness. Sources: George Grantham Bain Collection/Library of Congress, Washington, District of Columbia, American Institute of Physics/Science Photo Library and Dobson, G. M. B., *Applied Optics* 7 (1968), 387–405.

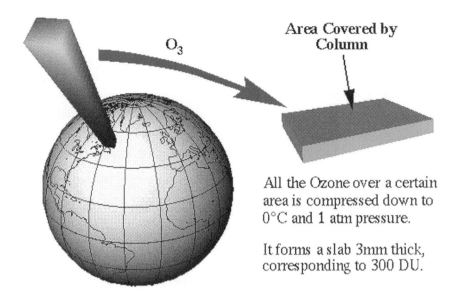

O_3

Area Covered by Column

All the Ozone over a certain area is compressed down to 0°C and 1 atm pressure.

It forms a slab 3mm thick, corresponding to 300 DU.

Figure 3.4. Definition of the ozone column and the units used for its characterization. DU stands for Dobson units. Source: https://community.windy.com/topic/7950/ozone-layer-and-ozone-hole-in-general.

is entirely absorbed in the atmosphere. They noted from their measurements made during the summer of 1912 by photographic photometry that, at this wavelength and for the Sun at the zenith, the proportion of radiation transmitted to the Earth's surface is only of the order of 1%. And in their paper published in 1913, they concluded that

> To produce this absorption, the atmosphere must contain an amount of ozone that is equivalent to a 5 millimeters layer of pure ozone under normal pressure.[3]

This was the first attempt to measure the amount of vertically integrated ozone with a value that is somewhat larger than currently accepted. This latter quantity, known as the *ozone column*, is an important physical quantity because it determines the total attenuation (from the top of the atmosphere to the Earth's surface) of solar ultraviolet radiation that passes through the atmosphere (Figure 3.4).

---

3. "Pour produire cette absorption, il faut que l'atmosphère contienne une quantité d'ozone équivalente à une couche de 5 millimètres d'ozone pur sous la pression normale."

Fabry and Buisson added that, if this quantity of ozone was uniformly distributed in the atmosphere, the proportion of ozone in the air would be equal to 0.6 cubic centimeter of ozone per cubic meter of air (corresponding to a volume mixing ratio of 600 ppb). Measurements made in the lower layers of the atmosphere by chemical method indicated, however, that the ozone proportion is about 0.008 cubic centimeter per cubic meter of air (corresponding to a volume mixing ratio of 8 ppb), thus a factor of 75 lower than can be explained by the measured absorption. Already in 1913, Fabry and Buisson concluded that

> The most probable hypothesis is that ozone is present only in the very high atmosphere where it would be produced by the extreme part of the solar ultraviolet radiation, which, being strongly absorbed by oxygen, can only act on the upper layers of the atmosphere.[4]

After the First World War, using their laboratory measurements of the ozone absorption, Fabry and Buisson were able to derive the ozone column with higher accuracy. They deduced this quantity from the accurate measurements at the Earth's surface of the solar radiation at about ten different wavelengths between 292 nm and 314 nm, that is, in the spectral region located between the Hartley and the Huggins bands, and in which absorption varies rapidly with wavelength. Based on the measurements made in Marseilles between May 21 and June 23, 1920, the two scientists determined with more certainty than in 1913 that the thickness of the ozone layer is of the order of 3 mm, with irregular variations ranging from 2.85 to 3.35 mm. They reemphasized that most of the ozone must be located in the upper layers of the atmosphere without knowing precisely at which altitudes.

In 1921, Fabry joined the Sorbonne in Paris as professor of General Physics and, from 1927 onwards, took on the Chair of Physics at Ecole Polytechnique, which was left vacant by Perot's death in 1925. During the same year, Fabry was elected member of the French Academy of Sciences and in 1931 became a foreign member of the Royal Society in London. His university courses attracted many students who were impressed by the clarity of his lectures and enjoyed the charm of his personality. Under the German occupation, Fabry stayed near Marseilles and returned to Paris in 1945, the year of his death.

---

4. "L'hypothèse la plus probable est que l'ozone existe seulement dans la très haute atmosphère, où il serait produit par la partie extrême du rayonnement ultra-violet solaire, qui, étant fortement absorbé par l'oxygène, ne peut agir que sur les couches les plus élevées de l'atmosphère."

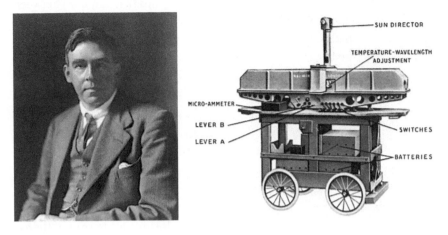

Figure 3.5. Gordon Dobson (left), who made systematic measurements of the ozone column with the spectrophotometer he developed at Oxford (right). Source: https://www. npg.org.uk/collections/search/person/mp76418/gordon-miller-bourne-dobson and https:// www2.physics.ox.ac.uk/research/atmospheric-oceanic-and-planetary-physics/history/ biography-dobson. See also Dobson, G. M. B., *Applied Optics* 7 (1968), 387–405.

After the work of Fabry and Buisson, a next important step during the 1920s was taken by Gordon Miller Bourne Dobson (1889–1975) in Oxford, UK. Dobson (Figure 3.5) had been hired in 1920 as a meteorology demonstrator at the Clarendon Laboratory of Oxford University by Frederick Alexander Lindemann[5] (who would become Lord and Viscount Cherwell), Winston Churchill's scientific advisor and confidant and a member of the Royal Society. Dobson had noticed that air temperature increases with altitude in the stratosphere. He suggested that the related warming of this atmospheric layer was likely due to the absorption of solar ultraviolet

---

5. Frederick A. Lindemann (1886–1957) studied aeronautical engineering in Berlin. He had been a talented pilot during the First World War. In spite of his German origin and accent, he was officially a British citizen since his father, an Elsassian engineer, had acquired the British citizenship. This intriguing personality, a member of the conservative party, and a vegetarian, who used the service of a driver to take care of his Rolls Royce and was known to be xenophobic, had since 1921 a close relation with Winston Churchill. He helped German scientists to flee from the Nazi regime and find research positions for them in the United Kingdom. Sometimes recognized as "the most powerful scientist ever," he was called "The Prof." by his supporters and "Lord Berlin" by his British opponents.

radiation by ozone. He also analyzed the relationships that may exist between weather fluctuations and the atmospheric column of ozone. To address these questions, he decided to systematically measure the amount of ozone and extended the first measurements made by Fabry and Buisson a few years earlier. With great ingenuity, Dobson built in 1924 a spectrograph (Figure 3.5) that recorded the solar spectrum on a photographic plate, and installed the instrument in a cabin built in the garden of his private property at Boars Hill near Oxford. The device, which had the advantage of being transportable, measured the spectrum of ultraviolet solar radiation. The ozone column was deduced from the ratio between the radiation intensities measured at two distinct but close wavelengths. The spectrum was produced by a curved-face prism developed around 1910 by the French physicist Charles Féry (1865–1935). A cell containing a mixture of chlorine and bromine filtered the radiation to retain only the ultraviolet wavelengths needed to measure ozone (Figure 3.6).

Two hundred days of observations carried out in 1925 revealed a very clear relationship between the meteorological situation and the measured value of the ozone column: this last quantity was highest in the vicinity of atmospheric low-pressure cells (cyclones) and minimum in high pressure zones (anticyclones). Observations performed the following year showed

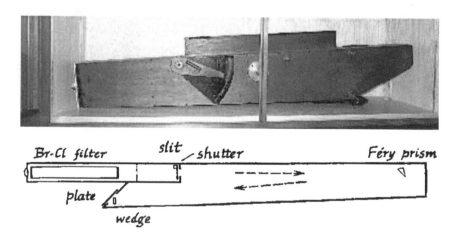

Figure 3.6. The first spectrograph built in 1924 by Dobson at Oxford and exhibited at the Clarendon Laboratory in Oxford, and a simplified schematic of the device. Almost 100 years after its invention and following some technological improvements, Dobson's device is still in operation in different areas of the world. Reprinted with permission from Dobson, G. M. B., *Applied Optics* 7 (1968), 387–405 © The Optical Society and http://www-atm.physics.ox.ac.uk/main/dept/dobson.html.

the existence of a previously unknown seasonal cycle in the ozone column. Dobson noted that, on the average, the ozone column at Oxford reached a maximum in spring (April) and a minimum in autumn (October).

To better understand the influence of weather conditions on ozone, Dobson decided to build five additional spectrographs, calibrate them to the original instrument, and install them first in several locations in Europe and later in different parts of the world. In Europe, the first measuring stations were located in Oxford in Great Britain, Valentia in Ireland, Lerwick in the Shetland archipelago of Scotland, Arosa in the Swiss Alps, Lindenberg near Berlin, Germany, and Abisko in Sweden. To obtain information on a global scale, some of the instruments were sent to Helwan in Egypt, Kodaikanal in India, Montezuma in Chile, Christchurch in New Zealand, and Table Mountain in California. In Europe, measurements were pursued systematically only in Oxford and Arosa. Further instruments were deployed later in other parts of the world.

One of the disadvantages of the method was the need to process a photographic plate for each measurement that recorded the intensity of the solar flux at different wavelengths. These plates were systematically sent to Oxford where more than 5,000 of them were analyzed and the values of the ozone column deduced. In order to make the data processing easier, Dobson decided to modify his instrument by replacing the photographic plate by a photoelectric cell that would directly provide the ratio of solar intensities at two distinct wavelengths. This modified instrument is generally called the Dobson spectrophotometer (Figure 3.5).

The measurements made by the Dobson instrument highlighted for each station the existence of a seasonal variation in the ozone column, already observed at Oxford, but with an amplitude that varied with latitude (Figure 3.7). This variation was small in tropical and subtropical regions (e. g., Chilean and Indian stations), but larger at high latitudes (e. g., Scottish and Swedish stations). Measurements also showed that the variation was offset by six months in the southern hemisphere compared to the variation observed in the northern hemisphere. Finally, on average, the amount of ozone in the atmosphere was significantly higher at high latitudes than in the tropics. Thus, there was every reason to believe that ozone is formed in the polar regions, perhaps in relation to the processes that generate the northern and southern lights above 100 km and known as aurora. If this hypothesis could have been verified—which it was not—one could have concluded that ozone diffuses downwards from the upper atmosphere. At this point in time, the question was clearly posed: how is ozone formed and destroyed in the atmosphere? Dobson's observations provided

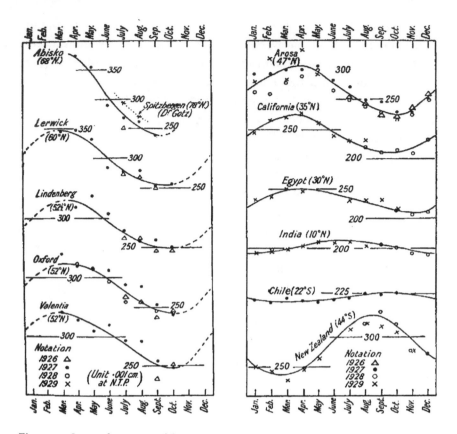

Figure 3.7. Seasonal variation of the ozone column expressed in Dobson units measured between 1926 and 1929 by several spectrographs installed in different regions of the world from 68°N to 44°S. These early observations revealed that the ozone abundance reaches a maximum in spring with the highest values found in the northern polar regions. They also highlighted the small amplitude of the seasonal variations in the tropics and a shift of six months in the seasonal variation in the southern hemisphere. Reproduced from Dobson et al., *Proc. Roy. Soc., London* A129 (1930), 411.

important constraints for those who wished to understand the chemical and dynamical mechanisms affecting ozone in the atmosphere. The chemists believed that the formation of ozone was intimately linked with the presence of oxygen, but the observations did not provide evidence that this was the case. Dobson, Harrison, and Lawrence wrote therefore in 1929:

> It has generally been supposed in the past that the ozone present in the upper atmosphere was formed from oxygen under the influence of the sun's ultra-violet radiation of wave-length about 1600 Å, but the results of the present

observations make it almost certain that this is not the chief cause of the formation of ozone. [...] It is more in accordance with the observations to suppose that the ozone is formed by some other cause and that the effect of sunlight is mainly to decompose the ozone already present.

A very large number of spectrophotometers—44 in 1956—were deployed around the world, and a global ozone monitoring network was established and administered after the Second World War by the International Ozone Commission (IO3C) created in 1948 in Oslo by the International Association of Meteorology and Atmospheric Sciences (IAMAS). Dobson was invited to chair the Commission and was assisted by a secretary, Sir Charles William Blyth[6] (1889–1982), who had recently stepped down as Director-General of India's Meteorological Service after the country became independent. It is interesting to note that one of the Dobson spectrophotometers was installed at the British Antarctic Base at Halley Bay during the 1957 to 1958 International Geophysical Year (see Box 1.1). The instrument observed a minimum of the ozone column during October and then a rapid increase in November. This phenomenon had no equivalent in the polar regions of the northern hemisphere. The Antarctic ozone minimum became considerably deeper after the 1970s, and very low ozone columns appearing each year in September and October were observed for the first time at the British polar station of Halley Bay and reported in the early 1980s.

During the Second World War, as the United Kingdom was conducting difficult bombing missions, the British authorities were concerned by the formation of visible condensation trails produced by the aircraft engines since these were revealing flight tracks to the enemy. Dobson worked with Alan Brewer (1915–2004), a member of the British Meteorological Office, to address this question (see Chapter 6 for more details). The collaboration continued after the war at Oxford University where Brewer developed a new

---

6. Sir Charles William Blyth Normand was a Scottish meteorologist educated in Edinburgh. He spent a long period of his life in India (1913–1944) as Imperial Meteorologist. He served in the military in Mesopotamia during the First World War and became Head of the India Meteorological Department (Poona) from 1927 to 1944 with the title of Director General of Observatories. He was among the founder of the India Science Academy. He returned to the United Kingdom in 1946 where he worked with Dobson on ozone measurements.

Figure 3.8. A Brewer spectrophotometer installed on the roof of the Steinbruch House in the ozone observatory located at Arosa in the Swiss Alps. (From Alois Feusi, Neue Züricher Zeitung, January 18, 2016.)

ozone sonde[7] and a spectrophotometer that still measure ozone today with a great accuracy (Figure 3.8).

Dobson retired from Oxford University in 1956 to pursue other passions, particularly music (he played the violin and the piano), gardening, and breeding. However, he continued his work on ozone from the station established in his Shotover Hill property. It was in his private observatory, financed by his own money, that he and his students carried out a large number of observations. He made his last measurement one day before being struck by a stroke and died six weeks later at the age of 86.

Starting in 1926, systematic measurements of the ozone column were also performed at an observing station installed near the quaint village of Arosa in the Swiss Alps. Astrophysicist Daniel Chalonge (1895–1977; Figure 3.11), a former student of Charles Fabry in Marseilles and one of the founders of the Institute of Astrophysics in Paris, installed in Arosa his spectrograph as well as part of the equipment built a few years earlier by Fabry and Buisson. Chalonge made extensive measurements during the day and the night, and

---

7. The ozone sonde developed by Brewer determines the ozone concentration as a function of altitude by measuring an electric current produced by the chemical reaction of potassium iodide in aqueous solution with atmospheric ozone molecules sampled outside the ozone sonde box by a small pump.

Figure 3.9. (Left) The Villa Firnelicht in Arosa constructed by Paul Götz as his private house. It accommodated the Lichtklimatisches Observatorium where measurements of solar radiation and of ozone were performed until 1953. Reproduced from Staehelin et al., *Atmos. Chem. Phys.* 18 (2018), 6567–84. (Right) Gertrud Perl, an Austrian scientist who moved to Switzerland and worked with Götz, makes ozone measurements with a Dobson spectrophotometer at the Arosa Observatory. Courtesy of J. Staehelin, ETH, Zurich.

showed that the thickness of the ozone layer does not vary according to whether the sun shines or not. The absence of a diurnal cycle in the ozone column was an important finding that constituted an additional constraint when explaining the origin of ozone in the atmosphere. Chalonge who was also an experienced mountaineer worked with the German scientist Friedrich Wilhelm Paul Götz (called Paul Götz, 1891–1954), who, after obtaining his doctorate at the University of Heidelberg, joined Switzerland. The researcher was well acquainted with the Alps, as he had tuberculosis and spent much of 1914 and 1915 in a sanatorium at Davos for treatment. He was convinced of the therapeutic qualities of clean air at high altitude. After arriving in Arosa in 1921, he founded the Lichtklimatisches Observatorium (LKO; Box 3.1) with the support of the local Chamber of Commerce of Arosa, the local tourist office, and of the Rhaetian Railway company.[8]

The observatory was created at a time when many were convinced of the benefits of solar therapy to cure tuberculosis, and therefore of the need to systematically measure insolation, particularly at high altitude. Götz developed the observatory by constructing with his own money a building which he called "Villa Firnelicht" (Figure 3.9). He deployed a series of

8. Private Swiss company established in 1888 by the Dutchman Willem Jan Holsboer and that operates trains mainly in the Grisons.

instruments on the site that measured solar radiation. He became interested in the processes responsible for the atmospheric absorption of solar light and installed a Dobson spectrophotometer, which from 1926 onwards continuously monitored the changes in the ozone column (Figure 3.10). Götz directed the station until 1953, shortly before his premature death a year later (see Box 3.1 for more details).

The observations continued after 1954, first by Götz's scientific Assistant Gertrud Anna Perl (1908–1974), who became the station manager in 1953 (Figure 3.9), then from 1967 to 1985 by Götz's former graduate student Hans-Ulrich Dütsch (1917–2003), who became a professor at the Zurich Polytechnic School (Figure 3.11). Dütsch had made extended scientific visits in the United States, first as a scientist at the High-Altitude Observatory (HAO) in Boulder, Colorado, and then at NASA near Washington, District of Columbia. The Arosa Observatory has played a major role in support of ozone research

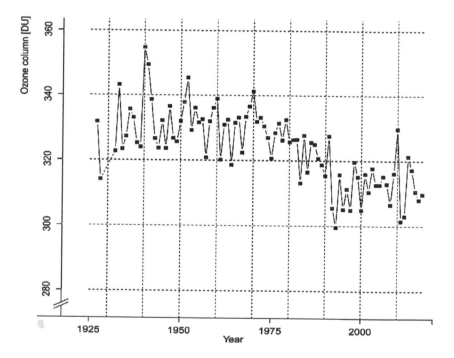

Figure 3.10. Evolution of the annual average of the ozone column (Dobson units) based on continuous measurements at the Alpine resort of Arosa in Switzerland since 1926. A filtering of the interannual variations shows that the amount of ozone in Arosa increased slightly between 1926 and 1955, and then decreased between 1955 and 2000. Reproduced from Staehelin et al., *Atmos. Chem. Phys.*, 18 (2018), 6567–84.

Figure 3.11. Hans Ulrich Dütsch (left, reproduced from Staehelin et al., 2018) and Daniel Chalonge (right, reproduced from https://www.noticiasdelcosmos.com/2011/08/medalla-chalonge-2011-john-mather.html) taking ozone measurements at the Arosa station in Switzerland.

and has generated the longest time series of the ozone column (Figure 3.10) that is currently available and that highlights long-term ozone trends and shorter-term variability.

In the late 1920s and early 1930s, networks of ozone stations were established in several countries on all continents including, for example, India. Measurements of total ozone were made in Kodaikanal (10°N) already during 1928 and 1929 and in Delhi (19°N) and Poona (18°N) during 1936 and 1938, using the photographic instrument developed by Dobson. Measurements based on the more elaborate Dobson spectrophotometer were initiated in 1940 by Kalpathi Ramakrishna Ramanathan (1893–1984) at the India Meteorological Department located in Poona with the encouragement and support of Sir Charles Normand, the Director General of the Observatories (Figure 3.13). Ramanathan, a fellow of the India Science Academy, was a senior scientist in this department from 1925 to 1948 before becoming the first director of the Physical Research Laboratory at Ahmadabad. Systematic measurements of total ozone were made, starting in 1951, at Ahmedabad, Kodaikanal, Sringar, and New Delhi.

Box 3.1. The Lichtklimatisches Observatorium of Arosa
(After Staehelin et al., 2018)

At the beginning of the twentieth century, before the discovery of antibiotics, tuberculosis was treated by placing patients in sanatoriums located in the mountains where the air is pure and the ultraviolet radiation is intense. In 1914 to 1915, the Deutsche Heilstätte Sanatorium in Davos, Switzerland, welcomed the young Paul Götz (Figure 3.12) who was coming from Germany to cure his pulmonary disease. Once in good health again, he taught at the German school of Davos before returning to Germany and in 1919 defended his doctoral thesis at the University of Heidelberg on the photometry of the lunar surface. In the years that followed, Götz returned to Davos to work with Carl Wilhelm Dorno (1865–1942), a wealthy industrialist from the German city of Königsberg (now Kaliningrad, Russia), who came to Davos in 1905 to accompany his daughter also affected by tuberculosis. Dorno (Figure 3.12), who had studied natural sciences in Halle and Königsberg and had completed his doctoral thesis in 1904, had started a research institute in 1907 to understand the importance of environmental factors in the treatment of tuberculosis. This institute, which he financed from his own means and directed until 1921, specialized in the measurement of solar radiation and was the precursor of the famous Physical Meteorological Observatory of Davos (PMOD), still in existence today. Dorno believed that intense solar radiation was good for curing tuberculosis.

Figure 3.12. Two pioneers of radiation research in the Swiss Alps: Carl Dorno (left, reproduced from Wikipedia), who created in 1907 the Physikalisch-Meteorologisches Observatorium in Davos, and Paul Götz (right, reproduced from Staehelin et al., 2018), who established in 1921 the Lichtklimatisches Observatorium in Arosa.

He published a large number of scientific papers and several scientific books, one on the physics of solar radiation and of sky radiation (Physik der Sonnen- und Himmelsstrahlung) and another one on the climatology in the service of medicine (Klimatologie im Dienste der Medizin).

Götz, whose relations with Dorno quickly became tense, decided to open a second solar radiation station, but this time in the tourist village of Arosa in the canton of Graubünden (Grisons), where several sanatoriums had also been established. He obtained funding from the local tourist office (the Kur- und Verkehrsverein or KVV), subsidized by tourist taxes levied on visitors to the alpine resort, which enabled him to supplement the meteorological measurements made in the village since 1884. Thanks to this funding, Götz was able to establish in 1921 the measurement station which soon became the "Lichtklimatisches Observatorium." Very quickly, he used measurements of solar ultraviolet radiation to deduce the integrated amount of ozone above the station. He was, for example, the first to notice that the value of the ozone column reaches a maximum in spring and a minimum in autumn. In 1926, the Observatory, first implanted in the property of a sanatorium of Arosa, settled in the villa Firnelicht that Götz built, and where the measuring station remained until 1953.

The Arosa Observatory attracted the best ozone specialists to make high-quality measurements. Götz welcomed two Oxford researchers, Gordon Dobson and his colleague A. R. Meetham, who installed a spectrophotometer on the roof of the Villa to continuously measure the amount of atmospheric ozone. The first attempts to determine the vertical distribution of ozone (from the Umkehr effect described in Chapter 5) were carried out in 1932 jointly by Götz, Dobson, and Meetham.

Götz, who had been appointed professor of meteorology at the University of Zurich in 1940, continued the ozone measurements in a systematic way despite the reduction of local funding by the KVV as a result of the economic difficulties caused by the Second World War, and despite the opportunities offered to Götz to go and work in a German university. With the emergence of antibiotics to treat tuberculosis after the Second World War and the resulting gradual closure of sanatoriums, the need to continue the research begun in Davos and Arosa became less obvious. Götz was able, however, to convince his funders that it was important in a health resort and winter sports station like Arosa to measure solar radiation and air quality. As a result, he was able to develop a project financed by the Swiss Transport Office and named "Klimaaktion." Ground-based ozone measurements were made with the assumption that high concentration values were beneficial for the health of the guests and other visitors to the alpine resort.

After Götz's premature death in 1954, the station was transferred from Villa Firnelicht, still occupied by Götz's wife, Margarete Karoline Beverstorff, to the Arosa Florentinum sanatorium. After a few months of interruption, the measurements were continued under the responsibility of Gertrud Perl. Perl, a native of Austria, who had completed her doctoral dissertation in 1935 at the University of Vienna at a time where only few women were enrolled in universities. Four years later, apparently to escape the Nazi regime since her family probably had Jewish roots, she moved to Switzerland, lived in Davos between 1941 and 1948 before joining the Arosa Observatory, where she became Götz's scientific assistant. She became the head of the station in 1953, but had to leave Arosa in 1962 because of serious health problems.

Thanks to the new funding from the KVV, the municipality of Arosa and the Canton of Graubünden, which coincided with the appointment in 1965 of Hans-Ulrich Dütsch as a professor at the Polytechnic School of Zurich, a new exciting period began for the Arosa Observatory. Dütsch continued the measurements of total ozone and adopted the Umkehr method to derive information about the vertical distribution of ozone in the stratosphere. He also initiated ozone measurements in altitude using light sondes suspended beneath small balloons. The ozone measurements made jointly at Arosa and the neighboring stations of Payerne in Switzerland and Hohenpeissenberg in Bavaria proved very useful for understanding the meteorology of the stratosphere. After the 1970s, the Arosa observatory became a reference station to quantify the ozone depletion produced by the emissions of chlorinated products generated by the industry, and in particular to calibrate the space-borne instruments used to measure the ozone column. In 1974, the Observatory left the Florentinum and moved to the Steinbruch house a few hundred meters away, in an environment with better working conditions. After the retirement of Dütsch in 1985, the Lichtklimatisches Observatorium of Arosa was transferred to MeteoSwiss, which continues the systematic measurements of ozone. Thanks to the perseverance of Swiss researchers, especially Götz, Perl, and Dütsch, the measurements made in Arosa, which began in 1926, constitute the longest ozone time series available today.

Ramanathan, who introduced and supported ozone research in his country, worked together with Ragnath N. Kulkarni on the evaluation of Umkehr observations made in India and elsewhere in the world (see Chapter 5 for a description of the Umkehr effect). He also studied the relations between the observed ozone distribution and the atmospheric

Figure 3.13. K. R. Ramanathan (left panel reproduced from http://www.rri.res.in) and Sir Charles Normand (right panel reproduced from Bojkov, 2012) initiated an ozone program in India in the late 1920s and early 1930s. They installed several ozone spectrophotometers in the country. The systematic measurements made at Kodaikanal, Delhi, and Poona provided unique information about the vertical ozone profiles in tropical regions.

circulation. Kulkarni left India in the early 1960s to join the CSIRO Division of Meteorological Physics in Aspendale, Australia.

The first ozone measurements carried out in China were made by the French Jesuit Father Pierre Lejay (1898–1958). The Society of Jesuits under the direction of Father Augustin Colombel (1833–1905) had established in Zi Ka Wei near Shanghai, an advanced meteorological, magnetic, and astronomical observatory. The main building of the observatory, now a museum, is located in the compound occupied by the current Shanghai Meteorological Service.

Pierre Lejay (Figure 3.14), after completing his studies at the École Supérieure d'Electricité in Paris, obtained his doctor's title of mathematics in 1926. During the preparation of his thesis, he pursued theology studies and was ordained priest the same year. The young Jesuit immediately left to Shanghai where he settled until 1939. The Reverend who was interested in geodesy and upper atmosphere physics, occupied from 1930 the position of director of the Zi Ka Wei Observatory. In the early 1930s, he received on loan the No. 4

Figure 3.14. Aleksandr Khr. Khrgian, who established the national ozone network in the Soviet Union. Photo credit: Department of Physics, University of Moscow. (Right) Father Pierre Lejay, S.J. (with the Chinese name of Yan Yue-fei), a prominent French Jesuit and geophysicist who was the director of the Zi Ka Wei Observatory in Shanghai from 1930, and made the first ozone measurements in China in 1932. Photo Credit: Shanghai Meteorological Museum.

spectrograph built by Dobson and made measurements of the ozone column from 1932 in the astronomical facility of Zo Sé located on a hill 25 km south west of Zi Ka Wei and, at that time, little affected by the air pollution produced in the urban area of Shanghai. These observations showed that the quantities of ozone, generally ranging from 220 and 260 Dobson units, were somewhat lower than those observed in Europe. They revealed, however, that values were higher during anticyclonic periods when Siberian air is transported southwards. In a paper published in 1939 in the Bulletin of the American Meteorological Society, Father Lejay stated that air masses located over a measurement station often originate from other latitudes where the amount of ozone is actually different. And he became very supportive of the "transportation theory" introduced at the Ozone Conference held in Oxford a few years earlier. In other words, it became clear that local variations in ozone were due to the horizontal movements of air masses in the upper atmosphere rather than, as previously thought, to local variations in the atmospheric pressure. To test his hypothesis, Lejay proposed to establish several measurement stations along a latitude circle in China. He believed that successive propagating waves of ozone would be observed by such an observation network.

Father Lejay was elected correspondent of the Paris Academy of Sciences in 1935 and became a full member of the Academy in 1946. After returning to France in 1939, he was appointed Research Director at the CNRS in 1945. He held several important positions in Paris, and worked at the Observatory of Paris and at the Institut de Physique de Globe. He also led the French Ionospheric Office and was chosen in the 1950s to be president of the French Committee for the International Geophysical Year. He died in October 1958 from a stroke on the ship *Flandre* that brought him back to France from the general assembly meeting of the International Council of Scientific Unions (ICSU) in Washington, District of Columbia.

An important contribution to ozone research was made in the early 1930s by Ny Tsi-zé, a junior Chinese scientist, also known as Yan Jici (1901–1996), who, together with a colleague from Singapore, Choong Shin Piaw, provided early quantitative laboratory measurements of the absorption of light by ozone in the ultraviolet (215–340 nm). These data were crucial to perform optical measurements of ozone in the atmosphere and to calculate the photolytic destruction of this molecule. Ny Tsi-zé had completed his PhD in 1927 at the University of Paris under the supervision of Charles Fabry and was the first Chinese student who was awarded a doctor's title in France. In 1930, after returning to his country, Ny Tsi-zé cofounded the Chinese Chemical Society and taught at several universities. He is often considered to be the "father" of modern physics in China. After the foundation of the People's Republic of China in 1949, Ny Tsi-zé contributed to the development of the Chinese Academy of Sciences (CAS) and became the first director of the CAS Institute of Physics. He got involved in Chinese politics and was Vice-Chair of the National People's Congress between 1983 and 1993. Interestingly, most of his seven children became also prominent scientists (high-energy physics, social sciences) and engineers (digital computer and aircraft designers). One of them was beaten to death during the Cultural Revolution.

In the Soviet Union, ozone was measured since 1957 by the M-83 filter instrument using an optical pigmented glass with a broad spectral band pass. The principle behind this instrument is the same as for the Dobson spectrophotometer. However, initially, it used a broad band of more than 40 nm. Total ozone was derived from the measurement of the light intensity from the direct sun and from the zenith sky. A comparison of parallel observations made by this filter instrument and by a Dobson spectrophotomer was performed at the University of Sofia and reported by Rumen Bojkov, a scientist at the World Meteorological Organization in Geneva. He showed large systematic discrepancies (up to 30%) between the measured ozone values, specifically for large solar zenith angles and high concentrations of aerosols. As a result, the M-83 instrument was modernized in 1983 under the leadership of Genady Gustin, a scientist at the Main Geophysical Observatory in Leningrad, to become the more advanced M-124 device. Direct comparison with Dobson instruments showed that the relative error of the new filter instrument was considerably smaller; it did not exceed 3 to 5%. A national ozone network, initiated in the Soviet Union in the 1950s by Aleksandr Khristoforovich Khrgian (1910–1993; Figure 3.14), a professor for 50 years at the Moscow State University, included more than 40 filter instruments by the 1970s. This network was extended to other countries of the Soviet bloc and used to derive long-term trends of ozone in the eastern part of Europe.

Optical Measurements of Surface Ozone

The accurate measurements of ozone concentrations at the surface remained a real challenge for a long time, because the values derived from the color of the Schönbein test paper were strongly affected by the ambient humidity and by the presence of chemical pollutants in the air, specifically in urban areas. It was therefore suggested to derive near-surface ozone concentrations by measuring the absorption of ultraviolet light produced by ozone along a rather long path parallel to the Earth surface. The first measurements were made in 1929 by Fabry and Buisson, who installed their UV instrument in the French Provence and operated it for several months (October 1929–March 1930); they derived an average ozone concentration value of 22 ppbv. Additional attempts were made in the early 1930s. Götz and his colleague Rudolf Ladenburg (1882–1952)[9] from the Kaiser

---

9. Rudolf Walter Ladenburg (1882–1952): atomic physicist born in Kiel, Germany. To avoid the situation encountered by Jewish scientists under the Nazi regime, he moved in the early 1930s to the United States and became a research professor at Princeton University. After 1933, he was the principal coordinator for job placement for exiled physicists in the United States.

Wilhelm Institute of Physical Chemistry and Electrochemistry in Berlin-Dahlem attempted to measure the absorption by near-surface ozone of the ultraviolet light emitted by a mercury lamp installed on a building in Arosa (Figure 3.15). The quasi horizontal optical path length between the lamp and the recording spectrograph was 3.6 km at an altitude of 2,300 m. From the measurements of the absorption of the light at different wavelengths (254–275 nm) made on May 4, 1930, the total amount of ozone along the path could be derived and a mean concentration of 29 ppbv was deduced for Arosa. Values of the order of 20 ppbv were obtained from similar

Figure 3.15. Paul Götz next to the instrument used to measure the total ozone amount along a quasi-horizontal path near the surface between a source of ultraviolet light and the recording spectrograph. The instrument shown here is deployed on April 30, 1930, near Arosa. Photo credit: Staehelin et al., The value of Swiss long-term observations for international ozone research, in *From weather observations to atmospheric climate sciences in Switzerland*, ETH Zürich Research collection, 2016.

measurements made at other locations in Switzerland (Lauterbrunnen, Chur, Jungfraujoch, Zurich) and in France (Marseilles). The concentrations derived by the French ozone specialists, Daniel Barbier,[10] Daniel Challonge,[11] and Etienne Vassy,[12] in Abisko, Swedish Lappland during the winter 1934 and 1935 provided similar results. All these spectroscopic observations highlighted the day-to-day variations in the ozone concentrations and suggested that the ozone concentration increases with height.

---

10. Daniel Barbier (1907–1965): French astronomer born in Lyon, France.

11. Daniel Challonge (1895–1977): French astronomer and astrophysicist. Student of Charles Fabry.

12. Etienne Vassy (1905–1969): French geophysicist, specialist of the upper atmosphere and stratospheric ozone. He was interested in using rockets to explore the upper atmosphere and launched an instrument on the French rocket Véronique to study radio-electric properties of the atmosphere. Together with his wife Arlette Tournaire Vassy (1913–2000), he organized several field campaigns aimed at measuring atmospheric ozone. Arlette Vassy was "Maître de Recherches" at CNRS and after 1968 became professor of Physics and director of the Laboratory of Atmospheric Physics at the Sorbonne in Paris.

CHAPTER FOUR

The First Theoretical Studies

The First International Ozone Conference

At the end of the 1920s, Gordon Dobson decided to convene a meeting with his colleagues around the world who had contributed to the development of the international ozone measurement network. He asked Charles Fabry, now Professor at the Sorbonne, to organize the meeting in Paris in 1929. Fabry announced this event in the following terms:

> It was suggested by Mr. G. M. B. Dobson that it would be useful to bring together people who have a direct interest in the issue of ozone and atmospheric absorption [...]. This meeting will take place at the Physics Laboratory of the Faculty of Science in Paris (Sorbonne) from 15 to 17 May.

Thirty-three researchers, mostly spectroscopists and meteorologists, enthusiastically responded to Fabry's invitation, and twenty-seven of them made a scientific or technical presentation at the symposium. The meeting was chaired by Dobson, who urged the participants to set up a working group that could foreshadow the establishment of a standing committee recognized by international scientific organizations.

The presentations made at the Paris meeting triggered a great interest among the participants. Dobson's presentation, first of all, highlighted the relationship between the measured amount of ozone and the current meteorological (synoptic) situation. The lectures of Chalonge and Götz reported that no diurnal variation could be detected in the ozone column. The famous Norwegian meteorologist Vilhelm Bjerknes (1862–1951) examined the relationship between ozone and tropospheric air movements, and the Swedish meteorologist Anders Knutsson Ångström (1888–1981), son of the prominent physicist Knut Angström[1] examined the relationship between ozone and the stratosphere's climate.

All these presentations suggested that ozone is more sensitive to weather fluctuations than to variations in insolation, and hence that this gas behaves primarily as an inert atmospheric tracer. The latest observations confirmed that the amount of ozone is maximal at high latitudes and minimal in equatorial and tropical regions, which were constraining factors for those seeking to identify processes of atmospheric ozone formation and destruction.

The First Photochemical Theory of Ozone

One of the presentations at the symposium created a real surprise. British mathematician and geophysicist Sydney Chapman (1888–1970; Figure 4.1) who, at that time, worked at Imperial College in London and was already enjoying a great reputation among his colleagues, suggested that chemistry plays an important role in the budget of ozone. Chapman was known for his work on stochastic processes, in particular Markov chains, and on the kinetic theory of gases. He also worked on the interactions between the solar wind and the Earth's magnetic field, and was particularly interested in the processes that periodically produce these magnificent colored veils that can be seen in the skies of Scandinavia and Canada, known as northern lights or auroras. The audience probably expected Chapman to explain the high polar ozone concentrations observed by Dobson by invoking an ozone production mechanism related to auroral phenomena. Since ozone seems to be more abundant at high latitudes, it was indeed tempting to claim that this molecule, like auroras, is formed around 100 km altitude by the bombardment of energetic particles originating from the Sun. The molecule would then diffuse downward and reach to the lowest layers of the atmosphere.

---

1. Angström gave his name to a unit of measurement of wavelengths (1 Angström = 10^{-10} m).

Figure 4.1. Sydney Chapman who introduced the first chemical scheme to explain the presence of ozone in the atmosphere. (Credit: http://celebrating200years.noaa.gov.)

Chapman, however, believed that other processes were involved. He probably knew that Yves Rocard (1903–1992), a physicist at the Ecole Normale Supérieure in Paris,[2] had shown that the vertical flow associated with the atmospheric diffusion of ozone in a nitrogen gas was only of the order of 20 m/day at an altitude of 50 km, and is therefore too slow to effectively transport large quantities of ozone to the lower atmosphere. Chapman suggested that ozone is produced by the action of ultraviolet radiation provided by the Sun, and introduced five chemical reactions involving only oxygen compounds. These five reactions should explain the balance between the formation and destruction of ozone and atomic oxygen in the atmosphere (see Box 4.1).

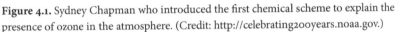

2. And the father of the former French Prime Minister, Michel Rocard (1930–2016).

> **Box 4.1. The Chapman chemical scheme (1929)**
>
> The five reactions proposed by Sydney Chapman at the Paris meeting in 1929 are as follows:
>
> $$O_2 + h\nu \rightarrow O + O$$
> $$O + O + M \rightarrow O_2 + M$$
> $$O + O_2 + M \rightarrow O_3 + M$$
> $$O_3 + h\nu \rightarrow O + O_2$$
> $$O + O_3 \rightarrow O_2 + O_2$$
>
> Ozone formation results from the recombination of the oxygen atom (O) with the oxygen molecule (O_2) in the presence of an inert molecule (M) that stabilizes the reaction product. The oxygen atom is produced by the photodissociation of the oxygen molecule by solar radiation whose incident energy exceeds the energy of the chemical bond of the O_2 molecule. Photodissociation of ozone produces an oxygen atom O, which recombines with O_2 to reform ozone, and therefore does not constitute a real loss of ozone. Ozone destruction occurs through the reaction between O and O_3. This reaction is relatively slow, especially at low temperatures. The recombination of atomic oxygen by the second reaction is important only in the upper atmosphere above about 80 km.

The first reaction that initiates the production of the ozone molecule is the dissociation of the oxygen molecule (O_2) into two oxygen atoms (O and O) under the effect of ultraviolet solar radiation. The oxygen molecule, which is very stable and hence very abundant in the atmosphere can only be destroyed by light if the incident photon carries sufficient energy to break the bond between the two oxygen atoms, that is, if the wavelength of the incident radiation is less than 242 nm. This radiation is only available in the upper layers of the atmosphere because it is completely absorbed by oxygen molecules long before it reaches the Earth's surface and even the troposphere. In the upper atmosphere, the oxygen atoms produced by this photochemical reaction recombine to reform the oxygen molecule, but this process can be ignored below 80 km altitude because another reaction, this time between the oxygen atom (O) and the oxygen molecule (O_2), produces the ozone molecule (O_3).

Ozone itself is subject to a process of dissociation by solar radiation, but the chemical stability of this molecule is considerably lower than that of oxygen, and the energy required to break the molecule is relatively low. The corresponding photodissociation threshold is therefore located in the near-infrared region (1180 nm), indicating that the ozone molecule can be broken into two fragments, O and O_2, by ultraviolet and even visible radiation, and thus can be dissociated in all atmospheric layers down to the Earth's surface. Finally, in his simple chemical scheme, Chapman showed that the net destruction of ozone results from the reaction of ozone (O_3) with atomic oxygen (O), which reproduces molecular oxygen (O_2). Further work to be discussed in the following chapters will show that this last, relatively slow destruction reaction can be accelerated (or catalyzed[3]) by the presence of other chemical species. But, this was not known at the time of the Paris symposium.

The chemical scheme presented by Chapman, which represents the first theory on the mechanisms for the formation and destruction of atmospheric ozone, appealed to participants at the Paris meeting because it explained for the first time why most of the ozone is confined to the upper layers of the atmosphere. The first calculations made by Chapman with the chemical rate constants as they were known at the time showed that the ozone layer should be located around 45 or 50 km altitude, which seemed to be in line with the observations made at the time, by Götz and Dobson. The calculations were based on an early laboratory determination of the absorption coefficients for molecular oxygen in the ultraviolet radiation. Some experimental data were available since the early 1900s; thanks to laboratory work carried out in Germany by Victor Schumann (1841–1913), a far-ultraviolet specialist, and by Carl David Tolmé Runge (1856–1927), well known to mathematicians as one of the authors of Runge-Kutta's numerical method for solving differential equations (Figure 4.2). Schumann had worked for a long time in a commercial company that was building industrial machines, which allowed him to accumulate enough money to build his personal spectroscope. This instrument allowed him to study the spectrum of hydrogen and oxygen molecules, usually in the evening after his working hours, during his leisure time. This led him to discover in the spectrum of oxygen between the wavelengths of 175 and 205 nm, the existence of a complex structure of bands superimposed on a continuum. The very intense absorption in this spectral region leads to the total extinction of shortwave solar radiation above the stratosphere. By adopting the

---

3. Catalysis is a process that increases the rate of chemical reaction by the presence of a given substance (catalyst) in much smaller quantities than the reactive species.

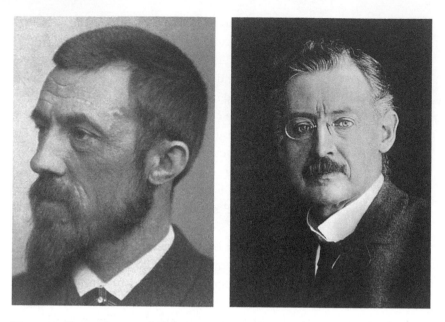

Figure 4.2. Victor Schumann (left) and Carl Runge (right) who observed the absorption bands of molecular oxygen in the ultraviolet radiation. Credit: https://de.wikipedia.org/wiki/Viktor_Schumann and https://commons.wikimedia.org/wiki/File:Voit_202_Karl_Runge.jpg.

absorption coefficient values measured at the time, it could be inferred that ozone is formed photochemically only above 50 km altitude, a conclusion that was modified 20 years later when additional measurements of the oxygen absorption spectrum became available.

The Following International Ozone Conferences

After the meeting held in Paris, the small "ozone community" decided to enhance its research activities in order to accelerate progress on what was considered to be a new and exciting research topic. A second ozone conference (Figure 4.3), this time with 58 participants, was organized in Oxford, UK, from September 9–11, 1936, at the invitation of G. Dobson. Twenty-nine papers were presented and addressed different questions: the methods for measuring ozone, the vertical distribution of ozone, the absorption of radiation and temperature of the upper atmosphere, as well as the relation between the horizontal distribution of ozone and weather conditions. Chapman discussed his photochemical theory of atmospheric ozone, Regener

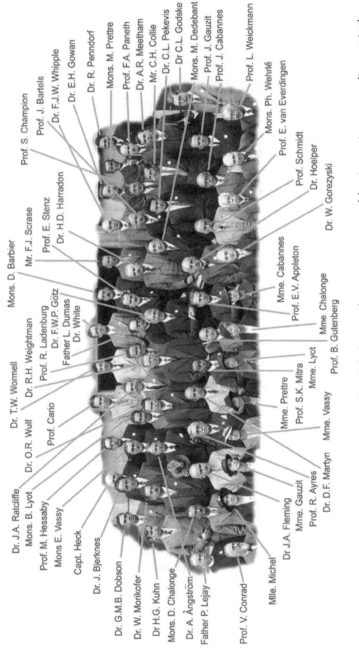

Figure 4.3. Participants to the Second Ozone Conference held in Oxford, September 9–11, 1936; many of the pioneers in ozone studies were in the attendance (courtesy of G. M. B. Dobson, Clarendon Lab., University of Oxford).

presented a paper on the ozone content of the stratosphere, and Götz made a presentation on the ozone time series recorded at Arosa. The papers by the different participants including Dobson, Erich and Victor Regener, Götz, and Chapman were published by the Royal Meteorological Society, several of them in English, and others in French or German.

With the outbreak of the Second World War, communication between scientists became difficult and no international meeting could be organized. Despite the war situation, however, a special meeting on the ozone problem was convened on April 17–18, 1944, in the small town of Tharandt near Dresden in Saxony, Germany. The meeting was hosted by the cloud physicist Helmut K. Weickmann (1915–1992) and involved the participation of Paul Götz, who came from Switzerland for the occasion. The proceedings, published a few years later (1949) by the German Weather Service (Deutsche Wetterdienst in der US-Zone), highlight the detailed discussions that took place about fundamental photochemistry, the distribution and the changes of ozone. In the following years, several ozone symposia were organized periodically in different regions of the world. Meetings[4] were held in Brussels (1951), Oxford (1952), Rome (1954), Ravensburg (1956), Arosa (1961 and 1972), Albuquerque (1964), Dresden (1976), Boulder (1980), Thessaloniki (1984), etc. The meeting in Ravensburg from June 25–29, 1956, was organized by Alfred Ehmert, a specialist of ozone sonde measurements at the Max Planck Institute for Physics of the Stratosphere in Weissenau near Ravensburg (Baden-Württemberg; Figure 4.4). Important discoveries on the photochemical destruction of ozone were reported at the two Arosa symposia. During the 1961 meeting, the "post-Chapman" ozone scheme involving the catalytic ozone destruction by reactive hydrogen radicals was presented and discussed. At the 1972 meeting convened by Hans-Ulrich Dütsch (Figure 4.5), the catalytic ozone destruction by nitrogen oxides was at the center of the deliberations. In the recent period, the Quadrennial Ozone Symposia have been organized by the International Ozone Commission (IO3C; Box 4.2); because of the high interest in the ozone question, each of the most recent meetings have attracted several hundreds of participants.

---

4. See http://www.io3c.org/sites/io3c.org/files/History_of_IO3C.pdf.

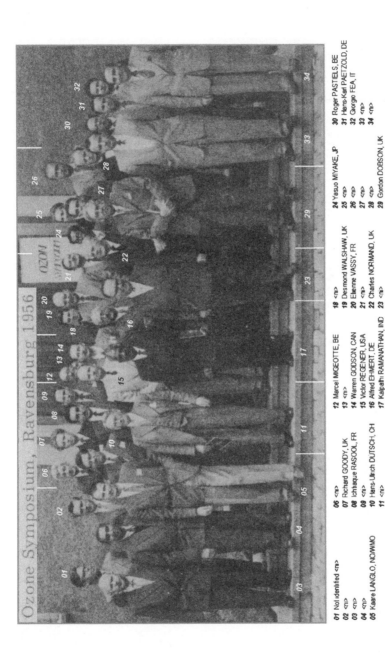

Figure 4.4. Participants of Ozone Symposium that took place in Ravensburg (Southern Germany) from June 25–29, 1956. Several participants including Gordon Dobson, Charles Normand, Etienne Vassy, Hans Dütsch, Victor Regener are identified at the bottom of the figure. Source: Dobson, G. M. B., *Applied Optics* 7 (1968), 387–405.

Figure 4.5. Participants of Ozone Symposium that took place in Arosa in August 1972. The list of most participants is identified at the bottom and at the top of the figure. (Reproduced from Bojkov, 2012.)

Box 4.2. Two international activities: The International Ozone Commission (IO3C) and the stratosphere-troposphere processes and their role in climate (SPARC) project

The International Ozone Commission (IO3C) was established in 1948 as one of the special commissions of the International Union of Geodesy and Geophysics, which represents the community of geophysical scientists around the world. The purpose of the IO3C is to help organize the study of ozone, including ground-based and satellite measurement programs and analyses of the atmospheric chemistry and dynamical processes affecting ozone. The first president of the IO3C was Dr. Gordon Dobson of Oxford University. IO3C organizes the Quadrennial Ozone Symposia (http://www.io3c.org/).

The Stratosphere-troposphere Processes And their Role in Climate (SPARC) Project, a core project of the World Climate Research Programme founded in 1992, promotes and coordinates cutting-edge international research activities on how chemical and physical processes in the atmosphere impact and are influenced by climate and climate change. SPARC's activities first focused on the role of the stratosphere (and specifically ozone) in climate. The Program now calls for a "Whole Atmosphere" approach around three themes: atmospheric dynamics and predictability, chemistry and climate, and long-term records for climate understanding. SPARC organizes open science conferences every few years (https://www.sparc-climate.org/).

CHAPTER FIVE

Determination of the Vertical Distribution of Ozone

A t the beginning of the 1930s, a quasi-consensus based on the observations of Götz and Dobson, and on the theory developed by Chapman, seemed to have emerged: the ozone layer was believed to be located at around 50 km altitude. However, it was soon realized that this statement was not correct. This false conclusion was reached because of the lack of accuracy in the experimental determination of the vertical distribution of ozone, and the incompleteness of the laboratory data on which the Chapman photochemical model was based.

The Umkehreffekt

Götz decided to tackle the problem of the vertical distribution of ozone. The stratosphere is not easily accessible and it became necessary to develop ingenious methods to investigate the upper atmosphere. The Director of the Lichtklimatisches Observatorium made an important discovery in the late 1920s that led to significant progress: By measuring at two distinct but close wavelengths, the intensity of solar radiation scattered from the zenith, he noticed a particular effect: a reversal of the ratio of light intensities in late afternoon when the Sun gets close to the horizon (Figure 5.1). This inversion, called by Götz the "Umkehreffekt," corresponds to a strengthening of

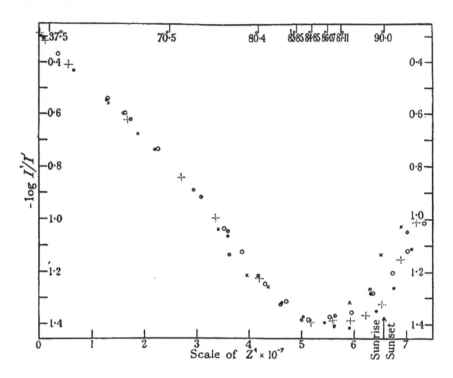

Figure 5.1. Logarithm of the ratio of radiation intensities from the zenith (blue sky) for two distinct wavelengths (311.4 and 332.4 nm) as a function of the Sun's zenithal angle. When this angle reaches 86.5° (or the Sun is 3.5° above the horizon), the curve bends and reverses. It defines the Umkehr effect noted by Götz in 1929. The different points correspond to observations made on different days in 1933. Reproduced from Götz et al. (1934).

the intensity of the ultraviolet radiation scattered by the atmosphere at the shortest wavelengths when the Sun approaches the horizon. Taking into account the spherical geometry of the Earth and its atmosphere, Götz, together with Dobson and A. R. Meetham, was able to infer from these observations the vertical distribution of ozone in the atmosphere in 1927.

Their measurements made in Arosa suggested that the ozone concentration reaches a maximum between 40 and 60 km altitude (Figure 5.2), a determination that was in agreement with the determination of the height of the ozone layer made in Marseilles in 1928 by Charles Fabry's former assistant, the physicist Jean Cabannes (1885–1959), a specialist in optics and by the astronomer Jean Dufay (1896–1967), a specialist in interstellar matter, both members of the French Academy of Sciences. Cabannes was a specialist of atmospheric optics and had been working after 1910 on the scattering of light by air molecules. Interestingly he married a daughter of Eugène Fabry,

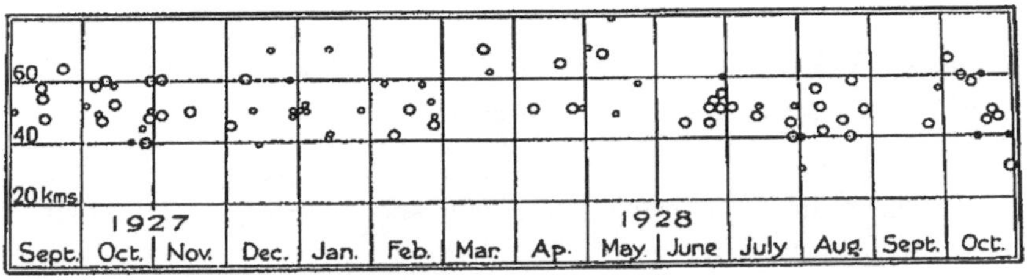

Figure 5.2. Altitude of the maximum ozone concentration derived by Paul Götz and Gordon Dobson in 1927 and 1928, using the Umkehr method. Reproduced from Götz, F. W. P. and G. M. B. Dobson, *Proc. Rot. Soc. London, Series A* 125 (1929), 292–94.

the brother of his supervisor Charles Fabry. Cabannes' research activities were interrupted during the First World War, but, in order to complete his thesis, he returned to Fabry's laboratory in 1919. The early estimates of the height of the ozone layer (approximately 50 km altitude) were shown to be incorrect, and were reconsidered in the following years on the basis of additional and more accurate measurements.

The correction resulted from the measurements made in 1929 and 1931. During these two years, Götz led a scientific expedition to Spitzbergen, where he could find the appropriate geographical conditions to make a large number of accurate measurements with the Sun frequently close to the horizon. The analysis of the measurements showed that the maximum concentration is located near 25 km altitude and not at 50 km, as was previously thought. Forty-six days of observations made in Arosa by Metham and Dobson during 1934 using the same technique indicated that the ozone maximum in Switzerland was located near 22 km altitude. The Umkehr method (Box 5.1) thus proved to be a very suitable technique for probing ozone in the atmosphere, even if the vertical resolution of the information it provided was limited. The method has been widely used in the following years to make systematic measurements of stratospheric ozone (Figure 5.3).

The measurements of the radiation scattered by the sky (Figure 5.4) led to another important discovery: the variations related to synoptic meteorological conditions detected by Dobson in the ozone column were in fact produced by ozone fluctuations in an altitude range of 10 to 20 km, and not higher in the stratosphere. It was therefore established for the first time that, in the lower stratosphere, ozone is more influenced by atmospheric dynamics and transport than by the photochemistry initiated by solar radiation.

The Umkehr method was increasingly recognized as a powerful technique and was adopted in different countries to derive the vertical profile

Box 5.1. The first measurements of the vertical distribution of ozone as reported by G. M. B. Dobson

"During 1927–29, an attempt was made to measure the average height of the ozone in the atmosphere by taking solar spectra with the Féry spectrograph over the maximum possible range of the height of the sun. [...] Many spectra suitable for these measurements were taken at Arosa with the Féry spectrograph and sent back to Oxford for development and measurement but the heights deduced were much too high, as were those of other workers using the same method.

In 1929 Götz took a Fabry-Buisson type of spectrograph, as well as the Féry spectrograph, to Spitzbergen to make measurements at very high latitude. His measurements made in July and August, confirmed the general increase in mean ozone value with increasing latitude and also showed the fall in ozone during the summer months owing to the annual variation. While in Spitzbergen, Götz took spectra of the zenith skylight while the sun was rising or setting and found that, when the sun was fairly high, the shorter wavelengths decreased in intensity more rapidly than longer wavelengths with increasing zenith distance of the sun. This was to be expected, owing to the greater absorption of the shorter wavelengths. He found, however, that when the sun was very low, the reverse occurred and the shorter wavelengths decreased in intensity more slowly than the longer ones. He rightly interpreted this as being due to the fact that, when the sun is high, most of the short wavelength zenith skylight is scattered from the direct solar beam below the ozone region, but when the sun is very low the absorption of direct sunlight by the ozone is so great that the amount of light scattered above the ozone region and which passes down to the instrument by the short vertical path becomes predominant. Near the end of 1930 he wrote to me telling me of the effect and suggested that it might be used as a means of estimating the vertical distribution of the ozone in the atmosphere. He also suggested that the ratio of two wavelengths, as measured by the photoelectric ozone instrument, should show a similar effect. Not really believing in Götz's suggestion, I made measurements on the zenith skylight on the first clear day early in January 1931, starting before sunrise, and was surprised to find that the dial readings really did increase at first as the sun rose, then became constant, and finally showed the normal decrease with the increasing height of the sun. This was the first umkehr curve to be obtained. I immediately wrote to Götz telling him of the success of his suggestion. Then followed a lot of work, both making measurements on the zenith skylight and also trying to work out a theory by which we might calculate the vertical distribution of the ozone. It was at this time that A. R. Meetham joined us and contributed greatly to the work both in making the observations and in developing methods for calculating the vertical distribution."

Reproduced from "Forty Years' Research on Atmospheric Ozone at Oxford: A History" by G. M. B. Dobson (1968).

Figure 5.3. First attempts to derive the vertical distribution of ozone in the Alps at Arosa in 1932. The instrument on the left, manipulated by G. Dobson, measures the light scattered by the atmosphere (at the zenith), while the second instrument on the right, under the responsibility of A. R. Meetham, measures the intensity of direct solar radiation. The most important result of this six-week measurement campaign is that the average height of the ozone layer is about 22 km, not 40 to 50 km, as previously thought. Reproduced from Dobson, G. M. B., *Applied Optics* 7 (1968), 387–405.

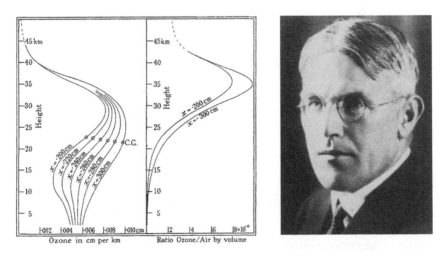

Figure 5.4. First determinations of the vertical distribution of atmospheric ozone (left, reproduced from Götz, F. W. P., A. R. Meetham, and G. M.B. Dobson, *Proceedings of the Royal Society of London. Series A* 145 (1934), 416) by Paul Götz (right).

of the stratospheric ozone concentration. In India, where an intense research activity had been initiated, Ramanathan worked together with Ragnath N. Kulkarni on the evaluation of Umkehr observations, and established the main features of the horizontal and vertical distribution of ozone in the tropics. Both scientists studied the relation between the observed atmospheric distribution of this chemical constituent and the atmospheric circulation. They extended their analysis to the extratropics, where they noted the large seasonal variation in the vertical distribution ozone (Figure 5.5) and the pronounced longitudinal differences in the abundance of this gas in late winter.

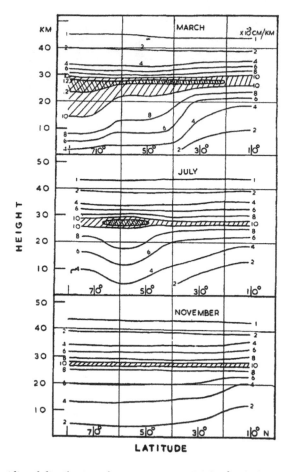

Figure 5.5. Meridional distribution of ozone concentration (10^{-3} cm of ozone per kilometer of air) in the Northern hemisphere for the months of March, July, and November performed by K. R. Ramanathan and R. N. Kulkarni (1960). From Ramanathan K. R., and R. N. Kulkarni, *Q. J. R. Met. Soc.* 86 (1960), 144–155, https://doi.org/10.1002/qj.49708636803.

The Pioneering Balloon Measurements

Götz's approach, although ingenious, had the disadvantage of being indirect. The German physicist Erich Rudolph Alexander Regener (1881–1955; Figure 5.6) decided therefore to carry out an in situ measurement of ozone in the stratosphere. Regener, who originated from West Prussia, joined the Technische Hochschule in Stuttgart where he became professor of Experimental Physics and Director of the Institute of Physics. He was interested in cosmic radiation discovered in 1912 by Victor Franz Hess (1883–1964) and, thanks to the electroscope he built, was able to measure with great accuracy the penetration of cosmic rays into the atmosphere and even into the water below the sea surface. This work allowed him to become familiar with the technique of hydrogen-filled balloons, which can carry heavy scientific payload into the upper layers of the atmosphere. Together with his son Victor H. Regener (1913–2006; Figure 5.6), he built a spectrograph capable of recording on a photographic plate light intensity in the ultraviolet at several wavelengths. A balloon carrying the instrument was launched from

Figure 5.6. Erich Regener (left) and Victor Regener (right) who determined the vertical distribution of ozone from the measured intensity of solar ultraviolet radiation from a stratospheric balloon. Source: Müller, R. (2009). A brief history of stratospheric ozone research. *Meteorologische Zeitschrift* 18, 10.1127/0941-2948/2009/353 and https://physics. unm.edu/pandaweb/history/faculty.php#Regener

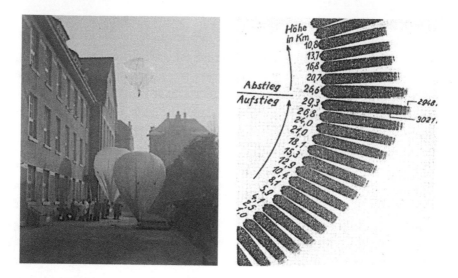

Figure 5.7. Launch of a stratospheric balloon at Stuttgart in 1934 (left panel) and recorded spectrum at several altitudes (expressed in kilometer) by photographic method during ascent (Aufstieg) and descent (Abstieg) of the instrument (right panel). Source: Regener, E., and V. Regener, *Physik. Zeitschr* 35 (1934), 788–93.

Stuttgart, Germany on July 31, 1934 (Figure 5.7) and reached an altitude of 34 km before exploding and letting the instrumental load slowly descend by parachute. The spectrograph recorded the ultraviolet spectrum of the Sun during ascent and descent. The photographic plate was recovered, and by applying Beer-Lambert's law[1] to ozone absorption at two distinct wavelengths, Regener deduced the vertical distribution of ozone concentration up to an altitude of 30 km.

Regener's interesting work was abruptly interrupted, however, in the autumn of 1937 when the professor from Stuttgart was suspended from his teaching duties by the National Socialist regime on the pretext that his wife Gertrud Heiter had Jewish ancestors. A few months later, Regener created the Stratospheric Physics Research Laboratory on the shores of Lake Constance, which was later integrated in the Kaiser Wilhelm Society and, after the Second World War, in the Max Planck Society. Erich Regener stayed in Germany during the world conflict, but his son Victor left Nazi Germany in 1938 to join the University of Padua in Italy, then the University of Chicago,

---

1. Relationship that links the attenuation of light to the properties of the medium it passes through and to the thickness traversed.

and finally the University of New Mexico where he completed his career as a professor in the Department of Physics and Astronomy.

One of the very dear wishes of Erich Regener was to continue his measurements in the upper atmosphere. Therefore, in 1939 he joined the Rocket Research Centre that the Nazi regime had established in 1937 in Peenemünde, along the Baltic Sea (Figure 5.8). Wernher von Braun (1912–1977), who was the director of the center, was interested in Regener's work. It is in these installations that the Third Reich developed and produced rockets for the war effort. The research center was bombed several times in 1943 and 1944 by the British and American aviation as part of the Crossbow operation to destroy the secret German weapon development. On August 17/18, 1943, 596 heavy bombers of the Royal Air Force took part in the first raid killing nearly 180 Germans and more than 500 foreign prisoners working in the facilities. Regener developed a spectrograph at Peenemünde to be launched by an Aggregat-4 rocket, also known as the V-2 rocket, to reach an altitude of 50 km and measure solar radiation up to this level. The Aggregat-4 rocket was the first guided ballistic missile which, in principle, was capable of carrying a load of one ton up to 250 km above sea level. Regener was able to perform one preliminary test, but the research program was interrupted in 1944 as the V-2 rockets were requisitioned to strike England and in particular to drop bombs on the city of London. The rocket payload that he had developed and

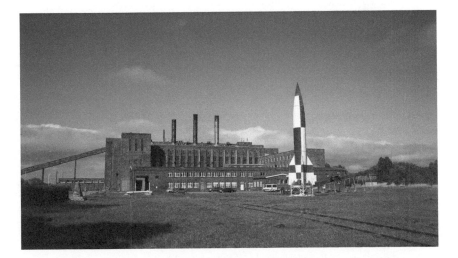

Figure 5.8. A V-2 rocket in front of the Rocket Research Center in Peenemünde, Germany near the Baltic Sea. Credit: Peenemünde Historical Technical Museum GmbH, https://museum-peenemuende.de/.

that the Germans called the "Regener-Tonne" suddenly disappeared from Peenemünde at the end of the war. A few months later, however, it was recovered in the United States. In 1945, Erich Regener was reinstated as professor at the University of Stuttgart and in 1948 became vice president of the Max Planck Society. In 1952, the research group led by Regener, then based near Ravensburg, moved to Lindau to join the Max Planck Institute for Stratospheric Physics. This latter institute merged in 1957 with the Max Planck Institute for Ionospheric Research founded by Walter Dieminger (1907–2000) to become the Max Planck Institute for Aeronomy and finally in 2004 the Max Planck Institute for Solar System Research. It is now based in Göttingen.

Balloon Flights with Manned Gondolas

In the early 1930s, the stratosphere was still an unexplored region at the frontier of knowledge. The Swiss physicist and explorer, Auguste Piccard (1884–1962; Figure 5.9), professor of physics first at the Federal Polytechnic School of Zurich and then at the Free University of Brussels (Université Libre de Bruxelles, ULB), had a dream: climbing in the upper atmosphere to measure the intensity of cosmic radiation. He designed therefore a vast balloon that should lift a sealed and pressurized gondola weighing several hundred kilos with two men on board. He invited his assistant Paul Kipfer (1905–1980) to join him in the gondola (Figure 5.9). The balloon was produced in Augsburg, Germany, and the gondola in Liège, Belgium. The launch was scheduled for September 14 1930, but the Germans tried to prohibit the flight because they judged the fabric of the balloon to be too thin. Piccard and Kipfer, who were Swiss citizens, were able to obtain permission to fly from the authorities of their country in Basel. A few minutes before the departure of the balloon, the German authorities reversed their earlier decision, but demanded that both explorers wear helmets during the ascent of the balloon. Marianne Piccard, Auguste's wife, who was present on the launch facility, quickly found the solution. She bought in a local store two wicker baskets that could be used as helmets, and also as stools for the two aeronauts while in the gondola. The weather rapidly deteriorated that day, so the launch had to be postponed for several months. On May 27, 1931, Piccard and Kipfer eventually embarked at Augsburg, Germany. The balloon filled with hydrogen was launched at 4 am and reached an altitude of 51,790 feet (15.781 km) before descending and landing 16 hours later on a glacier in the Tyrolian Alps close to the village of Gurgl. The flight was not without severe incidents, including a malfunction of the upper valve of the

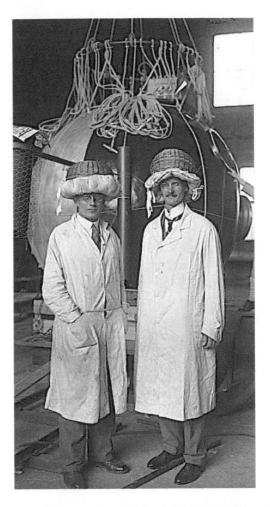

Figure 5.9. Auguste Picard (right) and Paul Kipfer (left) in front of the F.N.R.S gondola. The wicker helmets were designed to protect the two aeronauts at landing. They were also used as stools for the two explorers in the gondola while the balloon was flying. Credit: https://www.promi-geburtstage.de/info/?id=3380_Auguste-Piccard.

balloon, which did limit the ability of the crew to descend back to the surface. Piccard had named the balloon's gondola "*F.N.R.S*" because the expedition had been subsidized in Belgium by the National Fund for Scientific Research (Fonds National de la Recherche Scientifique) created three years earlier by King Albert I. A year later, in August 1932, Piccard repeated his flight during a second ascent, this time from the Dübendorf airfield near Zurich and with his assistant at ULB Max Cosyns (1906–1998). This time the balloon reached an altitude of 53,152 feet (16.201 m).

The expeditions had an extraordinary public impact at a time when different countries and specifically Belgium, the United States, the Soviet Union, Poland, and Spain were trying to capture the world's altitude record by manned balloons. Both flights organized by Piccard and his assistants set altitude records, but these were rapidly superseded by other stratospheric expeditions. For example, the Russian *C.C.C.P-1* expedition that took place on September 30, 1933, set a new record: the balloon reached 60,700 feet (18.501 m). Several successful flights designed to set altitude records were conducted in the United States by the Army, Navy, and Air Force. Since 2014, the world altitude record is held by Robert Alan Eustace, senior vice president of the Google Corporation who reached the height of 135.906 feet (41.424 m) using a balloon filled with helium. Eustace returned to Earth in a pressurized suit through a free parachute jump, and, as he descended, reached vertical velocities larger than 1,300 km/h.

In 1931 and 1932, Piccard, Kipfer, and Cosyns were celebrated as heroes in both Switzerland and Belgium. Hergé, the «father» of the well-known cartoon hero, Tintin, was directly inspired by Auguste Piccard when he conceived the famous professor Cuthbert Calculus (Figure 5.10) in his well-known albums.

After his journeys in the stratosphere, Auguste Piccard became interested in the ozone layer and, in a volume published in 1933 under the title *Au-dessus des nuages* (Above the clouds), he noted that, even though the mechanisms leading to the production of ozone in the upper atmosphere were not fully understood, it was important to remember that ozone absorbs solar ultraviolet radiation and allows therefore living organisms to exist at the Earth's surface. For Piccard, investigating the upper layers of the stratosphere was therefore essential. Rockets, he said, will be essential for scientific research and the only way to reach the limit of the atmosphere.

In 1934, Jean Piccard, the twin brother of Auguste, together with his wife Jeannette (Figure 5.11) ascended to more than 57,500 feet (17.5 km). Jeannette, a licensed balloon pilot, was the first woman to reach the stratosphere. She was an extraordinary woman. At the time of the flight, she had a bachelor's degree from Bryn Mawr College (1918) and a master's degree in organic chemistry from the University of Chicago (1919). She later earned a doctorate in education from the University of Minnesota. After Jean's death in 1963, she became a consultant for National Aeronautics and Space Administration (NASA). In 1974, she was ordained as an Episcopal Priest, which led to strong controversies because, at the time, priesthood was only accessible for men. In 1976, however, the Church decided to finally open priesthood to women, so that she could remain active in her duty. She passed away in 1981.

Figure 5.10. Professor Auguste Piccard has inspired Belgian cartoonist Hergé (Georges Remi) to characterize fictional character, Professor Cuthbert Calculus (Professor Tournesol in the French edition) in Tintin's albums. Copyright © Hergé/Moulinsart, 2019.

Figure 5.11. Jeannette Piccard was the first woman to enter the stratosphere. In 1934, she and her husband Jean Piccard (both seen in their gondola) ascended with the *Century of Progress* and reached the altitude of 57,979 feet. Credit: Museum of Science and Industry, Chicago. Source: https://www.nastarcenter.com/jeannette-piccard-first-woman-to-reach-the-stratosphere.

In the United States, the exploration of the stratosphere was also on the agenda. In 1933, US Navy Lt Cmdr Thomas G. W. «Rex» Settle and US Major Chester Fordrney reached the altitude of 61,000 feet (18.6 km) aboard the balloon «A Century of Progress». At the same time, United States Army Corp Captain Albert William Stevens, a First World War veteran, known for his courage and innovative thinking, tried to convince his superiors to mount a stratospheric expedition and make unique measurements of cosmic rays and ultraviolet radiation at high altitudes. The Army endorsed the project, but Stevens had to find the necessary funding. He contacted the National Geographic Society whose President Gilbert Grosvenor embraced the idea and provided most of the financial support. The Army Corp appointed a crew composed of Major William E. Kepner as the pilot and leader of the expedition, First Lieutenant Orvil Arson Anderson as an alternate pilot (and later the copilot) and Captain Stevens as a scientific observer. This team would fly in the pressurized magnesium alloy gondola built in Midland, Michigan, by the Dow Chemical Company, and measure the ultraviolet solar flux from which the vertical distribution of ozone in the atmosphere would be derived. Albert Stevens had become a photographer and a member of the US Army Signal Corps and later Chief of the Army Air Corps' Photography Laboratory in Ohio. His goal was to break the altitude record set by Auguste Piccard. The 3 million cubic feet balloon assembled by the Goodyear Zeppelin company in Akron, Ohio, and filled with hydrogen, took off on July 28, 1934, at 5:45 am above a canyon situated in the Black Hills National Forest (South Dakota) and in the presence of 30,000 spectators. Among them was Lorena McLain, the wife of the Governor of South Dakota Tom Berry and a large number of residents of the nearby Sioux reservation. The launch pad located 12 miles southwest of Rapid City, which became known as the Stratobowl, was a natural depression surrounded by cliffs that protected the balloon from wind gusts. The climb progressed without incident, but after seven hours of flight, as the aeronauts had reached the altitude of 63,000 feet and the balloon was fully inflated, the crew noticed cracks in the fabric of the balloon. Hydrogen was escaping rapidly through a large rip at the base of the balloon, and Explorer I began to descend with the three men on board (Box 5.2). Within 45 minutes, the balloon had reached the altitude of 45,000 feet, and 30 minutes later the height of 20,000 feet. The crew decided to dump ballast overboard including the heavy spectrograph that was measuring solar radiation. At 18,000 feet, the crew opened the door of the gondola. When the remains of the balloon (which played the role of a parachute during the descent) and the gondola reached an altitude of about 6,000 feet (2 km), the three men decided to

Box 5.2 The sequence of events following the collapse of Explorer I in 1934

At 1:15 pm, the 57,000 feet level was achieved when the crew heard a clattering noise overhead. Looking through the 3-foot upper port, and witnessed part of the appendix cord, falling on the roof of the gondola. As the crew looked upward, a 30-foot rip and three smaller tears in the lower part of the balloon were visible. As the balloon continued to climb to 60,613 feet (18,475 meters), the three aeronauts valved but the sun's rays expanded the hydrogen until the balloon was able to stop the ascent and begin downward 20 minutes later.

As Kepner readied to release the 80-foot parachute, the balloon envelope disintegrated. The temperature outside the gondola was 80° below 0° Fahrenheit while the inside temperature was just above freezing and getting colder as Explorer descended. Scientific instruments sounded warnings that enveloped the inside cabin, making it hard to talk between the three aeronauts who still needed to keep their routine of making oxygen, recording data, and deciding when to jump for safety.

In 45 minutes, Explorer was down to 40,000 feet, and then 20,000 feet another 30 minutes later. Both Kepner and Anderson forced open the hatch and scrambled out to take a look at the enormous bag that still held air in the upper section while the lower fabric wrestled to and fro from the wind speed, tearing it further apart. Suddenly the entire bottom of the bag dropped, allowing the top portion to act as a parachute. At this time, Kepner and Anderson cut loose the spectrograph, which floated to earth on its own parachute. It was time to "lighten the load" and slow the rate of descent. As Stevens jettisoned the remaining ballast, liquid air and their empty containers were fastened to parachutes and thrown overboard, lead ballast poured through the hopper in streams and through the hatch as each sack was opened individually.

At 10,000 feet the three still did not want to abandon the ship and continued on until Explorer reached 6,000 feet when Kepner jumped. Anderson was the second to grab the folds of his silk chute and exit the aircraft. At 5,000 feet above sea level or 2,000 feet above the ground in that part of Nebraska, Stevens was the last inside the gondola. As he prepared to exit, the remaining envelope burst into pieces and Explorer plummeted to earth. Twice Stevens tried to push himself out the spinning orb, only to have the wind pressure push him into the rapidly falling sphere.

Explorer was now at 3,500 feet when Stevens was able to propel himself out of the port-hole and into space. This may have been a fortunate thing, except now Stevens was falling next to and with the same speed as the gondola. As he pulled his rip cord, the chute expanded only to have a portion of the balloon fabric fall to the center of his canopy and then slide free. Stevens had enough time to look up and see the parachutes of Kepner and Anderson, then hear the tremendous thud of the gondola flattening onto the earth, before he too hit the ground seconds later.

The adventure of the first Explorer proved how important and dangerous the race to the stratosphere was for mortal men. Others who had tried before the success of Explorer II died in the race to become the first to witness the earth's aura from space.

From the Morning Star Balloon Company, http://www.nwplace.com/sbhistory.html.

jump using their emergency parachutes (Figure 5.12). Shortly afterwards, the hydrogen still included in the balloon exploded. Anderson standing on top of the rapidly falling gondola, jumped but had problems with his parachute that had collected pieces of the disintegrated balloon. Luckily, these pieces slid off and Anderson could land safely. Stevens left the gondola as it was in free fall only 150 m above the ground just a few seconds before the gondola crashed violently. The three aeronauts were safe, but most of the scientific data were lost. The spectrograph, however, which was cut loose by the crew just before the balloon exploded, came down smoothly by parachute and the photographic records were recovered intact on the ground.[2]

After this embarrassing failure, a second attempt was evoked. The National Geographic Society hesitated because in the meantime a stratospheric flight carried out in Russia also ended in failure, but this time with the death of the Soviet aeronauts.[3] Finally, it was decided that a 100,000 m^3 balloon inflated with helium would be produced by a specialized company and launched from the same canyon in the Black Hills. An attempt to take off with a new balloon, also assembled by Goodyear Zeppelin, was made on July 12, 1935, but an unexpected and brutal collapse of the balloon happened during its inflation, just before the gondola got attached to it. The helium escaped skyward and the fabric fell to the earth, hurting slightly

---

2. See for more details the article by Gregory Kennedy entitled "The two Explorer stratosphere balloon flights" (https://stratocat.com.ar/contacto-e.htm).

3. In fact, a total of seven Soviet aeronauts died on two high-altitude balloons in the early years of stratospheric exploration.

Figure 5.12. (Upper panel) Drawing representing the three aeronauts jumping by parachute after the explosion of Explorer I balloon. Credit: Tom Lovell/National Geographic Creative, http://www.pbs.org/wgbh/americanexperience/features/spacemen-balloon-innovation/. (Lower panel) The two aeronauts in front of the Explorer II gondola. Credit: Smithonian National Air and Space Museum, Washington, District of Columbia, https://airandspace.si. edu/collection-objects/lta-balloons-usa-explorer-ii-nov-1935-stevens-albert-william-capt-anderson-orvil.

some members of the supporting crew. The expedition was postponed until a technical committee would identify the cause of the sudden opening of the upper portion of the balloon.

The new attempt, under the name of Explorer II was scheduled to take place in October 1935. However, the weather was not favorable and the expedition had to be postponed for more than a month. During inflation of the balloon in the night of November 10, the fabric ripped 17 feet. Repair delayed by two hours the launch originally scheduled for 5:30 am. It was finally at 8 am on November 11, 1935, that a new balloon of 3.7 million cubic feet filled with helium, and a 640 pounds 9-foot spherical gondola carrying Albert Stevens (now leading the expedition) and Orvil Anderson (Figure 5.15) took off and eventually reached the world record high altitude of 72,395 feet (22 km), before successfully landing at about 4 pm near the small town of White Lake in South Dakota (Figures 5.12–5.14). This time, the harvest of scientific information was important; in particular, the aeronauts measured solar radiation at several wavelengths, which allowed them to deduce the vertical distribution of ozone and to compare their observations with the measurements made by Erich Regener a year earlier. After the successful mission, Albert Stevens wrote,

> … the balance of the ozone in our atmosphere has a tremendous influence on life as we know it. If ozone were rare, we would be sunburned by a few minutes' exposure of sun. If ozone were more, then we would probably die for lack of essential vitamins […] There would be an enormous increase in bacterial growth […] fatal to human life.

Figure 5.13. (Left) Launch of the Explorer II balloon from the Black Hills National Forest in South Dakota. Source http://stratocat.com.ar/artics/explorer-e.htm. (Right) The inside of the gondola. Source: Morning Star Balloon Co., http://www.nwplace.com/sbhistory.html.

Figure 5.14. The Explorer II gondola exposed at the Air and Space Museum in Washington, DC.

Figure 5.15. From left to right: Captain Albert W. Stevens, Major William E. Kepner, and Captain Orvil Anderson before the flight of Explorer I. Only Steven and Anderson took part in the Explorer II flight. Reproduced from http://www.airforcemag.com/ MagazineArchive/Documents/1979/June%201979/0679anderson.pdf.

Interestingly, as the balloon was going down, Stevens dropped a device that would collect airborne spores down to an altitude of 36,000 feet. The analysis in the laboratory of these samples revealed for the first time the presence in the stratosphere of ten types of spores, bacteria, and fungi, and hence of a variety of living microorganisms.

In the late 1930s and early 1940s, additional measurements of the vertical ozone distribution were made from unmanned balloons (Box 5.3) carrying light (2.3 kg) ultraviolet radiometers and related equipment (Figure 5.16). Several balloons launched either from Flagstaff, Arizona, or from Beltville,

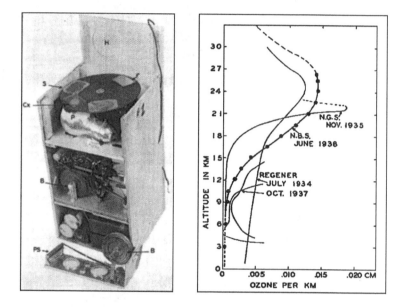

Figure 5.16. (Left) Ozone instrument used by Coblentz and Stair for different balloon ascents. The ozone column above the instrument is deduced from the transmission value by a calibrated filter of the incoming solar radiation. Light penetrates though a slit noted H in the lit of the instrument and reaches an optical cell located in the upper compartment together with a rotating disk. The middle compartment of the instrument hosts the motor and the lower part the radio system and the barometer. (Right) Vertical distribution of atmospheric ozone (cm column per km altitude) measured up to 30 km by Erich and Victor Regener in July 1934 by the Explorer II aeronauts (N.G.S.) between 0 and 22 km in November 1935 and by Coblentz and Stair (N.B.S) between 3 and 26 km in June 1938. The profile obtained in October 1937 by Victor Regener in the troposphere exhibits a decrease in the ozone concentration with height with a pronounced minimum at 5 km, and a gradual concentration increase at higher altitudes up to 12 km. The values provided by Regener appear to be overestimated in the lower troposphere. Reproduced from https://nvlpubs.nist.gov/nistpubs/jres/22/jresv22n5p573_A1b.pdf.

Figure 5.17. Anna Mani (right) who contributed substantially to ozone research in India talking to a colleague meteorologist. Credit: https://info.umkc.edu/unews/celebrating-women-in-stem-anna-mani/.

Maryland by W. W. Coblentz and R. Stair, both working for the National Bureau of Standards, reached heights of about 23 to 27 km and confirmed that ozone is localized in a layer located between 15 and 27 km with a maximum concentration near 24 or 25 km altitude.

Ozone research was also strongly supported in India by an interesting and charismatic personality: physicist Anna Mani (1918–2001). Mani (Figure 5.17), an expert in the development of meteorological instruments, put India at the forefront of ozone measurement programs. She undertook the development of an ozone-sonde and other instrumentation that enabled her country to collect reliable data for many years. As a young woman in the late 1930s, living in a society dominated by gender biases, Mani challenged her family and was allowed by her parents to enroll in a university. The brilliant and ambitious scholar, who was strongly influenced by Gandhian principles[4]—she was always wearing the *khadi*[5] as a symbol of her nationalist sympathies—joined Pachaiyappas College in Madras, where she earned in 1939 her bachelor

---

4. Mohandas Karamchand (Mahatma) Gandhi (1869–1948) was a leader of the Indian independence movement who promoted nonviolent civil disobedience.

5. The khadi, a handwoven natural fiber cloth, was worn by Indian women as part of an economic strategy to boycott foreign cloth. It represented an icon of the Swadeshi independence movement that aimed at removing the British Empire from power.

Box 5.3 The first sounding balloons: The discovery of the stratosphere

Georges Besançon (1866–1934) spent the first years of his life in the neighborhood of Pigalle in Paris before leaving the city with his mother in 1870 to settle in the suburban city of Colombes, less vulnerable to the bombardments that were common during the Prussian invasion of France. After completing his studies, Besançon, who was a strong advocate of the first French Republic, became a journalist and a popular science writer. In 1881, then aged 15, he met at the La Villette Park Wilfried de Fonvielle (1824–1914), a passionate aeronaut, a political writer, and a Republican activist, who introduced him to the techniques of ballooning. With several American citizens, Fonvielle had escaped in 1870 from occupied Paris aboard a balloon that took off under the fire of the Prussian guns.

In 1888, Besançon, who became prominent through his popular science papers, became a member of the Committee in charge of preparing the World's Fair (Exposition universelle) of Paris, which was planned to commemorate the centenary of the French Revolution of 1789. The same year, he founded the "Ecole Normale d 'Aérostation" which was established to train the future aeronauts, and was invited to direct the "Central Establishment of Aeronautical Constructions" in France. During this period, fascinated by air navigation, he made many flights in balloons and on July 11, 1892, while attempting to cross the English Channel, he almost lost his life when his balloon crashed in the sea.

In 1889, at the "Union Aérophile de France," Besançon met Gustave Hermite (1863–1914), a member of the French Academy of Sciences and the nephew of the brilliant French mathematician Charles Hermite (1822–1901). Hermite convinced Besançon to use balloons for scientific purposes. The two men then conceived an expedition to explore from the sky the Arctic regions, but the project was not financed and had therefore to be abandoned. From 1891 on, they launched a series of small balloons from an apartment located Boulevard Sebastopol in Paris to study their trajectory in the sky. Each balloon carried a postcard that had to be returned with an indication of where the balloon had been found. Starting in July 1892, Hermite and Besançon launched small balloons of oiled paper inflated with hydrogen from a city gas production plant at La Villette at the outskirts of Paris. In October 1892, they moved their operations to Noissy-le-Sec, and suspended to the balloons a barograph and a thermograph to record the air pressure and temperature during the ascent and descent phases of the flights. And so were born the sounding balloons that have been systematically used by meteorologists since 1898, and first by Léon Philippe Teisserenc de Bort (1855–1913) at the Meteorological Observatory of Trappes near Versailles.

Hermite and Besançon (Figure 5.18) repeated their experiments and made measurements up to 8,700 m altitude. In 1893, the two inventors decided to build a larger balloon, this time in rubber, which was expected to reach even higher altitudes: Aerophile I, launched on March 21, 1893, reached the level of

Figure 5.18 (Top left panel) Georges Besançon (on the right) and Gustave Hermite (on the left) in the gondola of a manned balloon carrying scientific instrumentation in 1897. Reproduced from http://radiosonde.eu/RS01/RS01B11.html. (Top right panel) The launch of the Aerophyle I carrying a barograph and a thermograph from the factory of Vaugirard on March 21, 1893. Reproduced from http://radiosonde. eu/RS01/RS01B20.html. (Bottom panel) A poster announcing that Besançon is running for the May 1898 parliamentary election in Paris. Reproduced from https://fr.wikipedia.org/wiki/Georges_Besan%C3%A7on.

16,000 m and highlighted for the first time a possible reversal in the vertical profile of the air temperature above 13,500 m altitude. Aerophile II, launched on October 20, 1895, recorded a temperature close to −70°C at the height of 11,000 m. In 1898, aeronaut Besançon, who enjoyed a great reputation among the population of Paris, became a candidate for the legislative elections in Paris (Figure 5.18).

Teisserenc de Bort, who belonged to a wealthy French family and had financed the construction of his Observatory in Trappes, was investigating dynamical processes in the atmosphere and was therefore interested by temperature measurements made at high altitude. He gathered his first data from instruments carried by kites connected to the ground by thin piano wires. With the achievements of Hermite and Besançon, he was able to use lacquered paper balloons filled by hydrogen and hence make measurements up to 20 km altitude. In 1898, he noted the presence of an "isothermal layer" above 10 km altitude, which confirmed the earlier observations from Aerophile I.

Teisserenc de Bort was not entirely convinced that this observation was not an instrument's artifact: the thermograph was likely heated by solar radiation, and he decided therefore to make nighttime measurements. He also decided to ignore the measurements above 10 km and extrapolate to high altitudes the

Figure 5.19 (Left panel): Richard Assmann (left) talking to Arthur Berson at the Lindenberg Observatory in 1907 (credit: https://en.wikipedia.org/wiki/Richard_Assmann). (Middle panel): Launch of the balloon filled with hydrogen and called *Preussen* at Berlin-Tempelhof on January 31, 1901. The balloon carrying Berson and Süring became the first manned balloon to reach the altitude of 10,000 m. From *Illustrirte Aeronautische Mitteilungen*, Heft 4, October 1901. (Right panel): Richard Süring (credit: Stefan Brönnimann and Alexander Stickler, *Meteorologische Zeitschrift*, 22(3) (2013), 349–58. See also Labitzke and van Loon, Berlin and the stratosphere, DOI https://doi.org/10.1007/978-3-642-58541-8 (Berlin Heidelberg: Springer-Verlag, 1999).

PHYSIQUE DU GLOBE. — *Variations de la température de l'air libre dans la zone comprise entre 8^km et 13^km d'altitude.* Note de M. L. TEISSERENC DE BORT, présentée par M. E. Mascart.

« J'ai l'honneur de communiquer à l'Académie les résultats de la discussion des observations rapportées par 236 ballons-sondes lancés de l'Observatoire de Météorologie dynamique et ayant dépassé l'altitude de 11^km, sur lesquels 74 ont atteint 14^km. Ces documents portent sur plusieurs années et sont répartis sur les diverses saisons.

» Ces observations, permettant d'étudier pour la première fois la température dans la zone comprise au-dessus de 10^km, mettent en lumière des faits nouveaux et imprévus dont le plus saillant est le suivant :

» 1° Alors qu'en moyenne la décroissance de température avec la hauteur augmente à partir des couches basses, et atteint dans les régions déjà explorées une valeur assez voisine de celle qui correspond aux variations adiabatiques de l'air sec, cette décroissance, au lieu de se maintenir à mesure que l'on s'élève, comme on l'avait supposé, passe par un maximum, puis diminue rapidement, pour devenir à peu près nulle à une altitude qui est, en moyenne, dans nos régions, de 11^km.

» 2° A partir d'une hauteur variable avec la situation atmosphérique (de 8^km à 12^km), commence une zone caractérisée par là très faible décroissance de température ou même par une croissance légère avec des alternatives de refroidissement et d'échauffement. Nous ne pouvons préciser l'épaisseur de cette zone ; mais, d'après les observations actuelles, elle paraît atteindre au moins plusieurs kilomètres.

Über die Existenz eines wärmeren Luftstromes in der Höhe von 10 bis 15^km.

Von Prof. Dr. RICHARD ASSMANN
in Berlin.

(Vorgelegt von Hrn. VON BEZOLD.)

Die neueren Forschungen über die verticale Vertheilung der Temperatur in der Atmosphäre haben den Beweis erbracht, dass die von JAMES GLAISHER gefundene schrittweise Verminderung der Abnahme derselben mit wachsender Erhebung auf einem grundsätzlichen Fehler seiner Methoden und Instrumente beruhte: es ergab sich im Gegentheil im Allgemeinen eine Vergrösserung des thermischen Gradienten mit der Höhe, wie es den Gesetzen der mechanischen Wärmetheorie entspricht.

Ausserdem aber erkannte man eine ausgesprochene Schichtenbildung im Luftmeere, die in engen Beziehungen mit der Wolkenbildung und mit horizontalen sowie verticalen Luftströmen steht.

Die bis zur Höhe von 9000^m reichenden directen Beobachtungen im bemannten Freiballon ergaben nach den Darstellungen der HH. BERSON und SÜRING im dritten Bande des Berichtswerkes »Wissenschaftliche Luftfahrten« vier durch ihre Eigenthümlichkeiten der Temperatur, Feuchtigkeit und Bewegung wohl charakterisirte Luftschichten, deren oberste durch nahezu adiabatisches Temperaturgefälle, geringen Wasserdampfgehalt und beträchtliche Windgeschwindigkeit ausgezeichnet ist.

Figure 5.20. Papers by Leon Teisserenc de Bort (top) published by the French Academy of Sciences in Paris and by Richard Assmann (bottom) by the Royal Prussian Academy of Sciences in Berlin, highlighting in 1902 the discovery of a warm layer above 10 km altitude. Credit: http://birner.atmos.colostate.edu/tropopause.html.

vertical temperature gradient observed below this 10 km. However, he eventually became convinced that the temperature was constant with height in the upper atmosphere and in 1902 published his conclusions in the *Comptes-rendus* of the French Academy of Sciences. He also coined the word *stratosphere* to define a yet unexplored region of the atmosphere, which appeared to be very stable.

The thermal structure of the upper atmosphere was also the subject of intense research in Berlin under the leadership of Richard Assmann (Figure 5.19) at the Prussian Meteorological Institute. On January 31, 1901, two meteorologists, Arthur Josef Stanislaus Berson (1859–1942), a native of Neu Sandez (Nowy Sacz) in Galicia and an assistant of Assmann as well as Richard Süring (1866–1950), a meteorologist native of Hamburg, took off from Berlin-Tempelhof in an open gondola balloon to measure the vertical temperature profile. The balloon reached the altitude of 10,800 m, but already at 6,000 m, the aeronauts felt that they were lacking oxygen, and at 10,000 m they became unconscious. They regain consciousness and successfully landed the balloon after 7.5 flight hours. The information gathered by Süring and Berson (Figure 5.19) was consistent with the measurements made from unmanned sounding balloons launched simultaneously from the ground.

Assmann concluded that there was therefore no reason to further distrust the measurements made from sounding balloons and became convinced of the existence of a "temperature inversion" above 10 km altitude. His conclusions were published just a few days after the publication of the paper by Teisserenc de Bort (Figure. 5.20). The French and Prussian studies conducted independently led to the discovery of the stratosphere.

In 1905, Assmann became the director of the Lindenberg Aeronautical Observatory in Beeskow and in 1909, Süring became the director of the Meteorologisch-Magnetisch Observatorium in Potsdam. Süring remained in this position until his retirement in 1932, but was called back in this job between 1945 and 1950 under the Soviet occupation.

Assmann, R., Über die Existenz eines wärmeren Luftstromes in der Höhe von 10 bis 15 km. (On the existence of a warmer airflow at heights from 10 to 15 km). Sitzber. Königl. Preuss. Akad. Wiss. Berlin, 24, 495–504, 1902.

Hartmann, G.: Georges Besançon (1866-1934), Perpétuel sinon Immortel. (Perpetual if not Immortal).

Teisserenc de Bort, L. P., Variations de la temperature de l'air libre dans la zone comprise entre 8 et 13 km d'altitude, Compt. Rend, Séances de l'Acad. Sci., Paris, 134, 987-989, 1902.

Photo credits:

degree in physics and chemistry. Mani then joined the Indian Institute of Science in Bengaluru where she worked on her doctoral thesis under the supervision of Nobel Laureate C. V. Raman (1888–1970). She submitted her dissertation to Madras University, but this academic institution refused to grant her the title of doctor with the excuse that she had not beforehand completed a master degree. Her dissertation, however, is still available today in the library of the Indian Institute of Science. She then spent a few years at Imperial College in London before joining in 1948 the India Meteorological Department located in Poona (today Pune), where she became a few years later deputy director-general and eventually member of the Indian National Science Academy and of the International Ozone Commission.

Ozone Measurements from Rockets

After the Second World War, the US military authorized the use of the V-2 rockets captured in Germany and transported to the United States to explore the upper atmosphere. A first rocket with a small automatic spectrograph was launched from the White Sands base in New Mexico on October 23, 1946. During the ascent, the camera photographed the solar spectrum from which Francis S. Johnson and his colleagues J. Dewitt Purcell and Richard Tousey of the US Naval Research Laboratory in Washington, District of Columbia, deduced the vertical distribution of the ozone concentration (Figure 5.21). Another measurement by the same type of rocket was made on April 2, 1948. The altitude at which the ozone concentration reached a maximum was found to be at 23.5 km during the first flight and at 18.5 km during the second experiment. These experiments provided a vertical profile of the ozone concentration only up to 38 km altitude. To determine the ozone distribution at higher altitudes, a small Aerobee rocket (8 m) was therefore launched from White Sands on June 14, 1949, at the time of sunset (7 pm). Under these conditions, the atmospheric path traversed by the solar ultraviolet light to reach the spectrometer was considerably longer than in the previous experiments, and the concentration of ozone could be derived up to 70 km altitude. In the 1960s, additional measurements were made by using small ARCAS (American Sounding System) rockets, also launched from White Sands. In these experiments, a probe developed by Victor Regener[6] was ejected

---

6. The method is based on measuring the light emitted by the reaction of ozone with a dye (Rhodamine B) contained in the instrument.

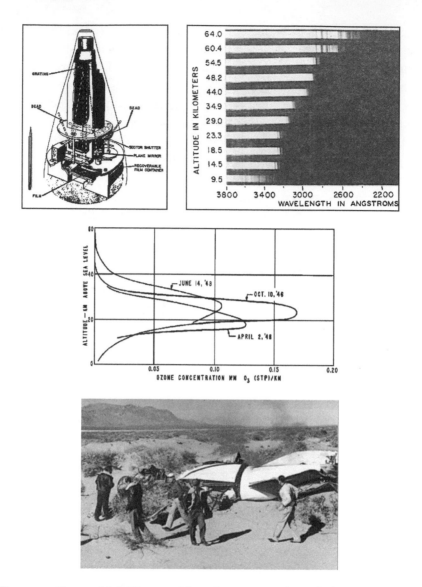

Figure 5.21. Top panel, left: Diagram of the rocket spectrograph installed in the tip of a German V-2 rocket and launched from the White Sands Proving Grounds, New Mexico. Top Panel right: Solar spectrum at different altitudes (9.5–64 km) measured by the spectrograph and recorded on a photographic plate on June 14, 1949 over New Mexico. Middle panel: Vertical distribution of ozone derived from the measurements made from two rockets on October 10, 1946, April 2, 1948, and June 14, 1949, respectively. Bottom panel: US Naval Research Laboratory scientists recover instruments from a V-2 rocket that has landed in the New Mexico desert (V-2 number 54 launched on January 18, 1951). Figures reproduced from Johnson et al. (1951), American Geophysical Union and https://en.wikiversity.org/wiki/Portal:Radiation_astronomy/Image#/media/File:Sl2lab06.jpg.

from the rocket at the altitude of 50 km. The instrument measured solar radiation during its descent, which was slowed down by the opening of a parachute. These measurements were supplemented by regular observations of ozone below 35-km altitude, made from balloons equipped either with the Regener instrument or with an electrochemical measurement system. These observations showed that the ozone concentrations calculated on the basis of the Chapman model overestimate by a factor of two to three the values actually observed. It became increasingly clear that the ozone theory had to be modified.

Satellite Measurements

After the launch of Sputnik by the Soviet Union on October 4, 1957 and the United States' effort to quickly close the gap in the space race, a question emerged: can stratospheric ozone be measured from satellites? It was a professor at the University of California at Los Angeles (UCLA), Sekharipuram V. Venkateswaran (1926–2005), who made a first attempt in 1960. He derived the vertical distribution of ozone by measuring the intensity of ultraviolet sunlight passing through the Earth's limb and reflected by one of the first communication satellites, Iota I. This study confirmed the presence of an ozone maximum at an altitude of about 25 km, but the measurements also suggested the existence of a secondary maximum at an altitude of about 55 km. The measurements taken two years later by R. D. Rawcliffe, a scientist working at the Aerospace Corporation, using a radiometer on board of an American Air Force satellite moving along a polar orbit, confirmed F. S. Johnson's earlier observations made from rockets; Rawcliffe did not detect the secondary maximum reported by Venkateswaran. Eleven years later, Paul B. Hays of the University of Michigan and Raymond R. Roble of the National Center for Atmospheric Research (NCAR) in Boulder, used an *occultation method*[7] to measure the atmospheric attenuation of light from several stars, and discovered the presence of a secondary maximum of ozone at high latitudes in the mesosphere, but this time around 85 km altitude rather than 55 km.

---

7. A method that allows a satellite to measure the absorption spectrum of the radiation emitted by the Sun or by a star and which passes through the atmosphere (through the limb) to reach a space probe. The abundance of chemical species in the atmosphere is deduced from this measurement.

Figure 5.22. Carlton L. Mateer, a Canadian scientist who made major contributions on the numerical processing of the Umkehr profiles, and the theoretical design (with J. V. Dave) of the BUV method to derive vertical ozone profiles from space.

In the mid-1950s, Austrian-born physicist Siegfried Frederick "Fred" Singer along with R. C. Wentworth suggested that stratospheric ozone profiles could be derived from the measurement by a downward looking spaceborne detector of the ultraviolet radiation backscattered by the atmosphere. In 1967, J. V. Dave and Carlton L. Mateer (1926–2011, Figure 5.22), who worked at NCAR in Boulder showed theoretically that the ozone column could be retrieved from such satellite measurements in the wavelength region of 312.5 to 317.5 nm. Their study provided the theoretical basis for the development of new spaceborne instruments, and in 1965, three separate attempts were made to measure ozone by this yet untested method. They were related to different space missions: the USSR Kosmos Mission (1964–1965), a US Air Force satellite (1966), and the NASA's Orbiting Geophysical Observatory (OGO, 1967–1969). In the Soviet Union, using the radiance measurements by the Kosmos-45 spacecraft, V. A. Iozenas, V. A. Krasnopol'skiy, and A. P. Kuznetsov were the first to derive the vertical ozone profile from measurements in the ultraviolet using a nadir-looking wavelength scanning monochromator. In 1966, a successful attempt was made in the United States by R. C. Rawcliffe, D. D. Elliot, M. A. Clark, and R. D. Hudson from the radiance measured at 284 nm by a polar-orbiting Air Force satellite. In 1967, additional measurements were performed during a 17-month period from the OGO-4 satellite and were analyzed by Gail Anderson, Charles A. Barth, F. Cayla, and Julius London at the University of Colorado in Boulder. The backscatter UV (BUV) experiment on Nimbus 4 was launched by NASA in April 1970 under the responsibility of Donald F. Heath at the Goddard Space Flight Center, followed in 1978 by an improved Solar backscatter UV (SBUV) instrument, also developed by Heath. This last instrument together with the Total Ozone Mapping Spectrometer (TOMS) developed by Arlin J. Krueger at NASA/Goddard was launched on Nimbus 7 satellite in October 1978. Before moving to the United States, J. V. Dave, who played a key role in the development of the backscatter method, had worked at the Physical Research Laboratory (PRL) in Ahmedabad, India, where, under the supervision of K. R. Ramanathan, he had investigated the radiative properties of the light scattered by the sky. Mateer developed the theory supporting the Umkehr method and the processing of the measurements made by Dobson spectrophotometers. He closely collaborated with Hans Dütsch, while the Swiss scientist was working at NCAR in the early 1960s.

In the early 1970s, it was shown that stratospheric ozone could also be retrieved from the measurements of the infrared radiation emitted by the atmosphere. Information on the total ozone column could be derived from radiance measurements made in the nadir direction, while vertical profiles could be retrieved by viewing the Earth's limb. A first attempt using this latter method was performed by John Gille at NCAR who developed the Limb Radiance Inversion Radiometer (LRIR), which flew on Nimbus 6 in 1975 and 1976. A more advanced Limb Infrared Monitoring of the Stratosphere (LIMS) instrument developed under the leadership of Gille and James Russell at NASA/Langley was flown on Nimbus 7 to determine the spatial distribution (vertical and horizontal) of ozone and other chemical species (H_2O, HNO_3, NO_2) in the stratosphere.

These different instruments made it possible to establish the climatology of ozone and other chemical constituents in the atmosphere, and to estimate with great accuracy the short-term variability, seasonal variations, and long-term trends of these atmospheric constituents. TOMS confirmed, for example, the increase in the ozone column with latitude, and specifically the presence of an ozone maximum that is produced near the North Pole at the end of winter (February to April) and near the latitude of 60° in the southern hemisphere (September to November). The presence of a low ozone column above the South Pole was also observed during October; the values in the 1970s were considerably higher, however, than those observed in the following decades. These observations unambiguously confirmed Dobson's findings reported in the 1920s and obtained at that time with much less sophisticated technology (Figure 5.23).

After the pioneering initiatives described above, several other spacecrafts carrying more sophisticated instruments were launched by the NASA and by the European Space Agency (ESA). The NASA's Upper Atmosphere Research Satellite (UARS; Figure 5.24), for example, deployed by the Space Shuttle in September 1991, offered a unique opportunity to monitor the evolution and fluctuations of ozone and other chemical species, and to perform an integrated investigation of the chemistry and dynamics of the stratosphere through 10 complementary (and sometimes redundant) instruments. For example, the Halogen Occultation Experiment (HALOE) led by James Russell from NASA/Langley made measurements of ozone, water vapor, methane, nitrogen oxides, as well as halogenated acids. The Microwave Limb Sounder (MLS) developed by Joe Waters from the Jet Propulsion Laboratory in Pasadena, California, derived the vertical profiles of ozone, water vapor, nitric acid, and chlorine monoxide. The ESA launched in 2002 a large Environmental Satellite (ENVISAT; Figure 5.24), carrying

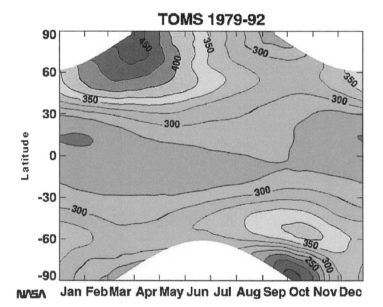

Figure 5.23. Mean values of the ozone column abundance (expressed in Dobson units) for the period 1979 to 1992. The zonally averaged values are represented as a function of latitude (from the South Pole to the North Pole) and the month of the year (from January to December). Credit: NASA.

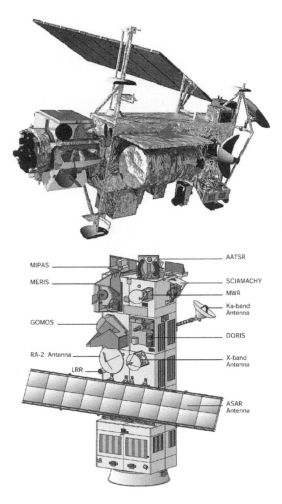

Figure 5.24. Upper panel: The Upper Atmosphere Research Satellite (UARS), a 5,900-kg NASA satellite developed to study the Earth's atmosphere and in particular the ozone layer. The spacecraft was deployed from Space Shuttle Discovery on September 15, 1991. The satellite with its ten instruments was revolving around the earth at an altitude of 600 km (370 mi) with an orbital inclination of 57°. The mission ended in June 2005, 14 years after the satellite's launch. A follow-up of UARS is the multinational AURA satellite dedicated to the study of ozone, air quality, and climate with specifically the Ozone Monitoring Instrument (OMI). It is the third major component of the NASA Earth Observing System (EOS) following on TERRA (launched 1999) and AQUA (launched 2002). Lower panel: The European Environmental Satellite (ENVISAT), the world's largest civilian Earth observation satellite (8,211 kg) with nine research instruments, was launched on March 1, 2002 from an Ariane 5 rocket into a Sun synchronous polar orbit at an altitude of 790 km (490 mi). The mission supported by the ESA ended on April 08, 2012, following the unexpected loss of contact with the satellite. It has been replaced since 2014 by the Sentinel series of satellites developed in Europe as part of the Copernicus program. Credit: https://directory.eoportal.org/web/eoportal/satellite-missions/u/ uars and https://earth.esa.int/documents/10174/669750/ESA_EO_programmes.pdf © ESA

nine instruments including the so-called Scanning Imaging Absorption Spectrometer for Atmospheric Chartography (SCIAMACHY) that provided quantitative information on a large number of chemical species in the stratosphere and troposphere by measuring the sunlight intensity transmitted, reflected, and scattered by the atmosphere in the ultraviolet, visible, and infrared. The Global Ozone Monitoring by Occultation of Stars (GOMOS), also on ENVISAT, measured the concentration of ozone, nitrogen oxides, water vapor, and aerosols in the stratosphere and mesosphere by detecting the atmospheric absorption of starlight in the ultraviolet, visible, and near-infrared. Several other instruments measuring ozone and related chemical species were launched more recently as part of the NASA Earth Observing System (EOS) and the ESA Copernicus Sentinel Programs. Some of these instruments such as the Tropospheric Monitoring Instrument (TROPOMI) launched in 2018 on the Sentinel 5 satellite, for example, provide tropospheric information on chemical species like nitrogen oxides at a horizontal resolution of a few kilometers, and allow the monitoring of air quality at the global scale. Infrared Atmospheric Sounding Interferometer (IASI) on board a European polar orbiting meteorological satellite (MetOp) provides measurements of a large number of chemical species in the troposphere.

In summary, satellites as well as Dobson spectrophotometers installed in different parts of the world and ozone sounding devices suspended under small balloons launched regularly by meteorological services have played an extraordinarily important role in establishing the climatology of stratospheric ozone. Today, space measurements of ozone and many other chemical species present in the atmosphere have moved to a large extent from frontier research activities to an operational mode. Satellites remain crucial platforms to quantify ozone depletion and other perturbations caused by human activity. We will return to this particular issue in later chapters.

CHAPTER SIX

Advances in Theory

In the late 1930s, the behavior of ozone in the stratosphere was relatively well characterized. It was known, for example, that the ozone layer is located around the altitude of 25 km and that the column abundance of this gas is considerably higher in the polar regions than in the tropics. It was also known that the ozone column is subjected to a seasonal variation: in the northern hemisphere, it reaches a maximum in April and a minimum in September. In addition, ozone in the lower stratosphere acts as a quasi-inert tracer, more sensitive to weather fluctuations than to photochemical effects. Ozone concentrations in the troposphere, especially near the ground, are significantly lower than in the stratosphere.

The First Mathematical Model of Stratospheric Ozone

The mechanism developed by Chapman and introduced at the Paris conference in 1929 was the only theoretical framework available at the time to explain the presence of ozone in the upper layers of the atmosphere. This mechanism implied that ozone, produced by solar ultraviolet radiation, is most abundant in the regions that are most illuminated by the Sun, that is, in tropical zones, and that its concentration outside the tropics is highest during the summer season. The opposite of what the observation showed. Chapman's theory, therefore, did not adequately explain Dobson's measurements. In addition, the theoretical calculations made at that time were uncertain because the kinetics parameters needed to estimate the ozone concentration were poorly quantified.

In 1931, Reinhard Mecke (1895–1969), a professor of physics and chemistry at the University of Heidelberg in Germany and a pioneer in infrared spectroscopy, attempted to calculate analytically the ozone concentration as a function of the atmospheric pressure (and therefore as a function of altitude), assuming photochemical equilibrium conditions in a "pure oxygen atmosphere" consistently with Chapman's theory. The question of the vertical ozone distribution was also addressed in 1936 and 1937 from a theoretical point of view by Oliver R. Wulf and Lola S. Deming (Figure 6.1), both working at the Bureau of Chemistry and Soils of the United States Department of Agriculture (USDA). As indicated in Chapter 3, Oliver Wulf (1897–1987; Figure 3.1) was primarily a laboratory

Figure 6.1. Lola Deming, a pioneer in ozone modeling, with her two daughters Diana and Linda in 1947. Reproduced from https://deming.org/deming/photo-gallery.

chemist. Lola Deming (1902–1986; Box 6.1) was a mathematician who moved from the USDA to the National Bureau of Standards in 1943 and worked there until 1963. She published several papers with her husband William Edwards Deming (1900–1993), a prominent engineer and an influential statistician who conducted an international career as a professor and a management consultant. At the time of Wulf and Deming's study, the balloon observations of Regener had just been released, and the two scientists were therefore aware that the maximum ozone concentration was located near 25 km altitude. Wulf and Deming decided to calculate the vertical distribution of ozone and developed what should be regarded as the first mathematical model of atmospheric ozone. They adopted the photochemical scheme proposed by Chapman, assumed for all reactive species equilibrium conditions and represented the solar spectrum by an idealized black body spectrum at the temperature of 6,000 K. A large uncertainty in the calculation resulted from the inaccurate derivation of the number of quanta absorbed in the atmosphere by molecular oxygen and ozone as a function of height because the absorption coefficients of these two molecules had never been measured in the spectral region around 200 nm. The authors of the study adopted therefore two rather arbitrary values for the spectral distribution of the absorption coefficients between the wavelengths of 210 and 235 nm (Figure 6.2). By solving the chemical kinetics equations, they derived therefore two possible vertical distributions of ozone between 15 and 40 km altitude. They also showed that the time for

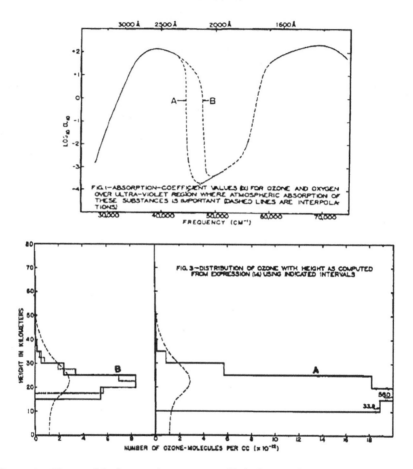

Figure 6.2. First model of stratospheric ozone published in 1937 by Oliver R. Wulf and Lola S. Deming on the basis of the chemical scheme presented by Sydney Chapman in 1929 at the first ozone conference held in Paris. (Upper panel) Absorption coefficients for ozone and molecular oxygen as a function of wavelengths and interpolation values (A and B) in spectral regions where laboratory data were not available. (Lower panel) Calculated vertical distribution of the ozone concentration (cm^{-3}) derived for cases A and B and compared to observational values (dashed lines). Reproduced from Wulf and Deming, *J. Geophys. Res.* 41 (1936), 299–310.

Figure 6.3. Gerhard Herzberg, laureate of the Nobel Prize for chemistry in 1971, who measured the absorption of ultraviolet radiation by the oxygen molecules at wavelengths larger than 200 nm. Credit: Government of Canada, National Research Council, https://depot-numerique-cnrc.canada.ca/eng/view/object/?id=7cb8b1ef-322f-4194-9456 86d52e5fca6d.

half restoration[1] of ozone equilibrium is of the order of minutes at 60 km, but is increasing toward lower altitudes to reach days at 30 km. In both model cases, the calculated ozone maximum was located between 20 and 25 km in agreement with the latest balloon measurements. In each simulation, however, the calculated ozone concentrations were considerably higher than the observed values (Figure 6.2). There was an urgent need to better characterize the absorption of ultraviolet light in the atmosphere, so that more accurate theoretical determination of the ozone abundance could be performed. Together with Reinhard Mecke, Lola Deming and Oliver Wulf should probably be considered as being the first "ozone modelers."

Ozone Production in the Stratosphere

The solution to this problem encountered by the pioneer ozone modelers was provided in 1950 by the future laureate of the Nobel Prize for Chemistry (1971) Gerhard Herzberg (1904–1999; Figure 6.3) together with

---

1. The half restoration time of a quantity (such as the concentration of a gas) is the time required for this quantity to return to half of its equilibrium value after it has been perturbed by an external event.

his "postdoctoral Fellow" from Germany, Peter Brix (1918–2007). Both identified a major source of stratospheric ozone that had been previously ignored. Herzberg, a native of Hamburg, Germany, studied engineering at the Technische Hochschule in Darmstadt where he completed his doctoral thesis. He then held various research positions, notably in Göttingen, but was forced to leave Nazi Germany in 1934 because his wife was of Jewish origin. He moved to Canada, taught at the University of Saskatchewan in Saskatoon, but it was at the National Research Council in Ottawa that he and Brix investigated the spectroscopic properties of the oxygen molecule. Brix, after his visit in Ottawa, worked at the University of Heidelberg and later joined the Technische Hochschule in Darmstadt before becoming Director at the Max Planck Institute for Nuclear Physics in Heidelberg.

As previously mentioned, it is the dissociation of the oxygen molecule (O_2) into two oxygen atoms (O and O) by ultraviolet solar radiation that leads to the formation of the ozone molecule (O_3). Herzberg and Brix completed the spectroscopic measurements made at the turn of the century by Schumann and Runge, who quantified the absorbing properties of the oxygen molecule at wavelengths shorter than 200 nm. The Ottawa group discovered that the molecule also absorbs ultraviolet radiation beyond 200 nm. And, since solar radiation at around 200 nm penetrates the atmosphere down to about 20 km, ozone is produced not only above 50 km, as previously thought, but also lower in the stratosphere. By applying these new laboratory data to Chapman's scheme, the calculated ozone concentration in the stratosphere reaches a maximum around the altitude of 25 km. In other words, the measurements made in Ottawa reconciled, but only in part, theory and observation.

The Inadequacies of Theory

The Chapman mechanism suggested that the presence of ozone in the atmosphere is entirely determined by the intensity of solar radiation, and that its atmospheric abundance should thus be largest in the tropics. The observations and more specifically the measurements made by Dobson showed, however, the opposite: the ozone column abundance is largest at high latitudes. The Chapman's mechanism did not explain either why the ozone column peaks at the end of winter or spring and not during summer. Despite these severe discrepancies, the chemical scheme led to an important step forward and even today, the simple chemical scheme proposed by Chapman remains the starting point for any investigation of stratospheric

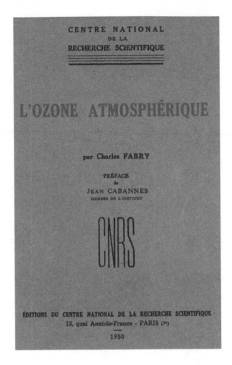

Figure 6.4. The book written by Charles Fabry in 1945, the year of his death, and completed five years later by Arlette Vassy, an ozone specialist at the Faculty of Sciences in Paris, describes very elegantly the state of knowledge on atmospheric ozone at that time. The book is prefaced by Jean Cabannes, a student of Fabry in Marseilles, who became a member of the French Academy of Sciences in 1949. Credit: Centre National de la Recherche Scientifique.

chemistry. However, new theoretical considerations had to be introduced to complement Chapman's ideas. Several prominent scientists were not at all convinced by the photochemical theory. In 1945, for example, Charles Fabry, in his book on atmospheric ozone (Figure 6.4), which curiously never quotes Chapman's work, wrote at the end of the monograph:

It is somewhat disappointing to note that after 30 years of research, we have not succeeded in constructing a coherent theory of atmospheric ozone [...] We have made a theory by admitting that solar ultraviolet is the only producing agent, with the long wave ultraviolet as a destructive agent, and we try to believe in the accuracy of this theory; but in fact, we succeed badly and we push the theory to the end almost without believing in its correctness. It is certain that the purely ultraviolet theory does not fit well with the facts observed in northern regions after the polar night. [...] Looking unbiasedly at the state of the matter, one gets the impression that we have not yet been able to pinpoint the crucial point of the problem, and that discoveries remain to be made, important not only for the theory of ozone, but perhaps also for general physics.[2]

Ozone Destruction by Hydrogenated Compounds

In 1950, an important step led to a first addition to the Chapman theory. The discovery was made by two European scientists who met in 1949 and 1950 at the California Institute of Technology in Pasadena, California: a mathematician and physicist from Northern Ireland, Sir David Bates (1916–1994) and a physicist and meteorologist from Belgium, Baron Marcel Nicolet (1912–1996; Figure 6.5).

Bates was an eminent specialist in molecular collision physics and was interested in the chemical composition of interstellar clouds and the physico-chemical processes occurring in the Earth's upper atmosphere. He spent much of his career at Queen's University in Belfast and became a Fellow of the Royal Society.

Nicolet, one of the "fathers" of aeronomy, the science that deals with the effects of solar radiation on the chemical species of the atmosphere, completed his doctoral thesis at the University of Liège on the chemical composition of stellar atmospheres. In 1935, he was hired at the Royal Meteorological Institute in Uccle by the then Director, Jules Jaumotte. During the Second World War, Nicolet worked on the process that leads to the formation of ionized layers in the upper atmosphere. These layers play an important role in the transmission of radio signals because they

---

2. "Il est un peu décevant de constater qu'après 30 ans de recherches, on ne soit pas arrivé à construire une théorie cohérente [...] de l'ozone atmosphérique. [...] On a fait une théorie en admettant que l'ultraviolet solaire est seul en jeu comme agent producteur, avec l'ultraviolet moyen comme agent destructeur, et l'on s'efforce de croire à l'exactitude de cette théorie ; mais en fait, on y réussit mal et l'on pousse la théorie jusqu'au bout presque sans croire à son exactitude. Il est certain que la théorie purement ultraviolette cadre mal avec les faits observés dans les régions septentrionales après la nuit polaire. [...] En regardant sans parti pris l'état de la question, on a l'impression que nous n'avons pas encore su mettre le doigt sur le point capital du problème, et que des découvertes restent à faire, importantes non seulement pour la théorie de l'ozone, mais peut-être aussi pour la physique générale."

Figure 6.5. Sir David Bates (left) and Baron Marcel Nicolet (right) who showed that water vapor destroys ozone in the upper atmosphere. Credit: https://qepgroup. wordpress.com/2017/02/13/early-career-researcher-nominations-sought-for-2017-bates-prize/ and http://www.uccle.be/administration/manifestations-publiques/images-2/ galerie-des-citoyens-dhonneur-de-la-commune-duccle/1987-baron-marcel-nicolet/view.

reflect electromagnetic waves. The Belgian scientist explained, for example, the origin of the ionosphere's D-region[3] at an altitude of about 80 km. In the 1950s and 1960s, Nicolet was particularly interested in space exploration and, by studying the drag of satellites by the upper air, he predicted the existence of a helium belt that surrounds the Earth above 700 km of altitude. The presence of this belt was later confirmed by Soviet space experiments. Nicolet became a foreign member of the French and US Academies of Sciences.

When they arrived in California in the early 1950s, Bates and Nicolet knew that Aden and Marjorie Meinel had observed the nocturnal emission of hydroxyl radical (OH) in the upper atmosphere and had detected their spectral bands, identified a few years earlier by Herzberg in the laboratory. Aden Meinel (1922–2011) had just completed his thesis at the University of California at Berkeley under the title: A Spectrographic Study of the Night Sky and Aurora in the near infrared. As will be seen later, the OH radical does not only produce what are now called the "Meinel bands," but it also plays a very important role in atmospheric chemistry because it is a powerful

---

3. The D-region of the ionosphere is an electron layer located between 60 and 90 km altitude and produced during the day primarily by the ionization of nitrogen oxide. It greatly attenuates the propagation of radio waves at high frequencies.

oxidant. It therefore destroys many chemical constituents of the atmosphere. If it is produced in an excited vibratory state, the relaxation of the radical to its fundamental energetic state is accompanied by the emission of light. Bates and Nicolet were also aware of the ozone observations made by spectrometers carried on board rockets launched by the Naval Research Laboratory in the United States. These measurements showed that ozone concentrations measured above 50 km altitude were significantly lower than those calculated using Chapman's theory.

Bates and Nicolet showed that photodissociation by solar radiation of water vapor molecules present above 70 km in the atmosphere produces hydroxyl radicals (OH) as well as hydrogen atoms (H) (Figure 6.6). The latter atoms react with ozone to produce hydroxyl radicals, while OH reacts with the oxygen atom (O) to reform atomic hydrogen (H).[4] During these two consecutive reactions, an ozone molecule (O_3) and an oxygen atom (O) are destroyed and transformed into two oxygen molecules (O_2). The net result of this mechanism is similar to the ozone destruction proposed by Chapman, but the action of hydrogenated compounds H and OH is to accelerate this destruction. These two reactions therefore constitute a catalytic cycle because the destructive agents, H and OH, are preserved by this process, which can therefore be repeated many times. When the Chapman mechanism is corrected by adding the reactions identified by Bates and Nicolet, calculated and measured ozone concentrations are in good agreement above 50 km of altitude.

The situation is different below 50 km altitude. John Hampson at the Canadian Armaments Research and Development Establishment demonstrated in 1964 that hydrogenated radicals contribute to the destruction of ozone in the stratosphere as well.[5] Two years later, an Australian scientist Barrie G. Hunt, who had just completed his PhD thesis at the University of Adelaide and was detached from the Weapons Research Establishment in Salisbury, South Australia, to the Geophysical Fluid Dynamics Laboratory (GFDL) in Princeton, New Jersey, conducted a detailed investigation of the ozone photochemistry in a so-called moist stratosphere. He concluded that, when adding the hydrogen reactions introduced by Bates and Nicolet and

---

4. The corresponding reactions are H + O_3 → OH + O_2 and OH + O → H + O_2. The net effect of this catalytic cycle is O_3 + O → $2O_2$.

5. In the stratosphere, the corresponding catalytic cycle results from the reactions: OH + O_3 → HO_2 + O2 followed by HO_2 + O_3 → OH + $2O_2$ with a net effect that is $2O_3$ → $3O_2$.

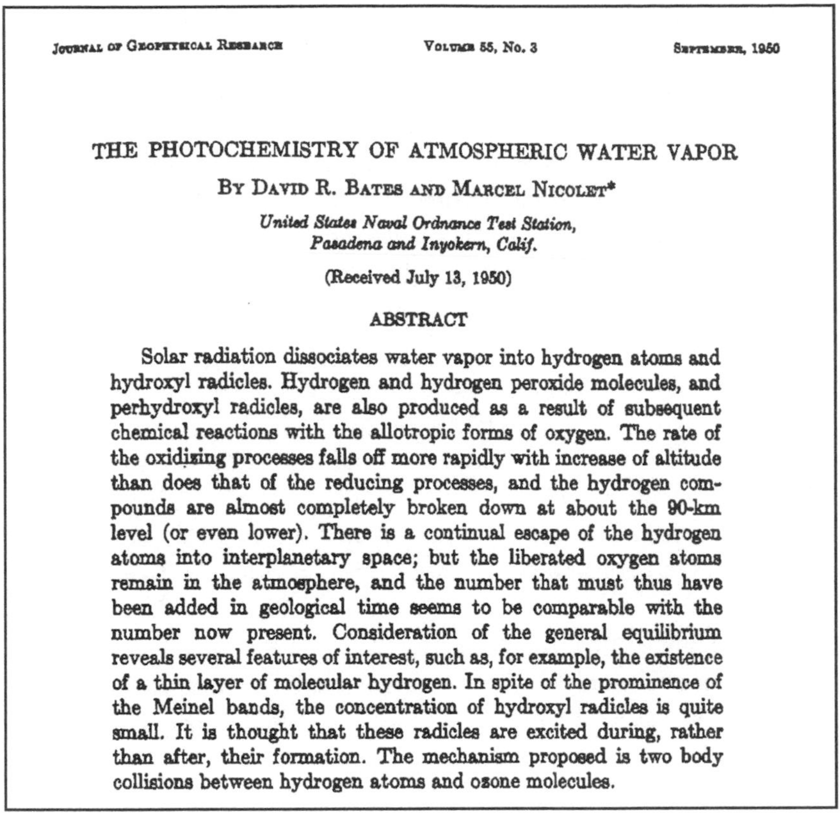

JOURNAL OF GEOPHYSICAL RESEARCH   VOLUME 55, No. 3   SEPTEMBER, 1950

THE PHOTOCHEMISTRY OF ATMOSPHERIC WATER VAPOR

BY DAVID R. BATES AND MARCEL NICOLET*

*United States Naval Ordnance Test Station,
Pasadena and Inyokern, Calif.*

(Received July 13, 1950)

ABSTRACT

Solar radiation dissociates water vapor into hydrogen atoms and hydroxyl radicles. Hydrogen and hydrogen peroxide molecules, and perhydroxyl radicles, are also produced as a result of subsequent chemical reactions with the allotropic forms of oxygen. The rate of the oxidizing processes falls off more rapidly with increase of altitude than does that of the reducing processes, and the hydrogen compounds are almost completely broken down at about the 90-km level (or even lower). There is a continual escape of the hydrogen atoms into interplanetary space; but the liberated oxygen atoms remain in the atmosphere, and the number that must thus have been added in geological time seems to be comparable with the number now present. Consideration of the general equilibrium reveals several features of interest, such as, for example, the existence of a thin layer of molecular hydrogen. In spite of the prominence of the Meinel bands, the concentration of hydroxyl radicles is quite small. It is thought that these radicles are excited during, rather than after, their formation. The mechanism proposed is two body collisions between hydrogen atoms and ozone molecules.

Figure 6.6. Summary of the paper on the photochemistry of water vapor published in 1950 by David Bates and Marcel Nicolet, while working at the California Institute of Technology and at the US Naval Ordnance Test Station.[6] Their paper described the hydrogen chemistry in the mesosphere and showed how the hydrogen chemistry affects ozone in the mesosphere. Credit: American Geophysical Union; reproduced from Bates, D. R. and M. Nicolet, The photochemistry of atmospheric water vapor, *Journal of Geophysical Research* 55 (1950), 301–27.

---

6. The primary function of the United States Naval Ordnance Test Station (NOTS) was the research, development, and testing of weapons. Starting in 1943, the station provided facilities and services to the California Institute of Technology (CIT) at Pasadena involved in the war effort. After 1945, the station became a permanent research and development center for the Navy. The main research and development facility is located near China Lake in the Mojave Desert 160 miles northeast of Los Angeles. Since 2014, it is now called Naval Air Weapons Station.

by Hampson in the simple chemical scheme introduced by Chapman, the theoretical results agreed satisfactorily with observation. This conclusion was questioned in 1970 by Nicolet who showed that some reaction rate constants adopted by Hunt in his calculation were considerably overestimated.[7] When the values of these rate constants were corrected, it clearly appeared that the ozone concentrations calculated in this region of the atmosphere were still a factor of two higher than observed. Clearly, another ozone depletion mechanism had not yet been identified and remained therefore to be discovered.

Ozone Destruction by Nitrogen Compounds

The missing mechanism was identified in 1970. The Dutchman Paul Crutzen (Figure 6.7) is credited with this achievement. Crutzen is an engaging and brilliant personality who, through his multitude of ideas, significantly advanced atmospheric and Earth system science in the last decades of the twentieth century.

With the limited financial means available to his family in Amsterdam after the Second World War, Crutzen was unable to undertake university studies. He decided, however, to enroll in a higher technical college from which he graduated as a civil engineer in 1954. His diploma allowed him to find a job at the Bridge Building Office in his city of Amsterdam. However, his dream and ambition were to work in a research institution. An advertisement published by Stockholm University, to which he responded, offered him the opportunity to start a second profession: he became a programmer and computer scientist at the Institute of Meteorology of this university. There, he was welcomed by Bert Bolin (1925–2007),

---

7. For example, Hunt adopted the value of 5×10^{-13} cm^3/s for the reaction between OH and ozone, which was the upper limit quoted in 1964 by chemist F. Kaufman at the University of Pittsburg in Pennsylvania. Nicolet indicated that a more appropriate value would be 5×10^{-15} cm^3/s at the tropopause (190 K) and 1.5×10^{-13} cm^3/s at the stratopause (270 K). Current laboratory estimates are of the order of 2.5×10^{-14} cm^3/s in the middle of the stratosphere (T = 220 K), and therefore substantially smaller than assumed by Hunt. A value of 10^{-14} cm^3/s was arbitrarily assigned by Hunt for the rate constant of the reaction between HO$_2$ and ozone. Nicolet stated that the role of such reaction in the stratosphere is doubtful due to the lack of experimental evidence for such chemical process. Current laboratory studies provide a rate constant for this reaction close to 1×10^{-15} cm^3/s in the stratosphere (T = 220 K), a factor ten smaller than assumed by Hunt.

Figure 6.7. Paul Crutzen, laureate of the 1995 Nobel Prize for chemistry, who discovered that nitrogen oxides destroy ozone in the stratosphere. Credit: European Parliament http://www.europarl.europa.eu/news/en/headlines/society/20110103STO11194.

a meteorologist, former student of Carl-Gustaf Rossby (1898–1957), and specialist of the carbon cycle and one of the initiators of the famous IPCC,[8] Henning Rodhe, who directed a program on acid rain and Georg Witt (1930–2014), a specialist in the noctilucent clouds that are present during the summer at an altitude of 85 km and produce magnificent colors in the polar sky. Crutzen contributed to the computer coding of weather forecasting models at a time when programming languages were still rudimentary and specific to each type of computer. Having noticed his exceptional aptitudes, the professors of the Institute of Meteorology suggested that he start a doctoral thesis for which he would develop a mathematical model describing the evolution of tropical cyclones. Crutzen was not very enthusiastic about the subject and decided instead to devote himself to a problem that was no longer in vogue in the 1960s: the chemistry of stratospheric ozone. With the passion of a self-taught person who has never had the

---

8. The Intergovernmental Panel on Climate Change (IPCC), established in 1988, provides periodic assessments of the state of scientific, technical, and socioeconomic knowledge of climate change, its causes, potential impacts, and response strategies.

opportunity to follow a full university curriculum on physics or atmospheric chemistry, he completed his thesis and continued his postdoctoral Fellow studies at the Clarendon Laboratory at Oxford University. It was there that he published an article in April 1970 that completely changed the scientists' view on stratospheric chemistry.

In Stockholm and then in Oxford, Crutzen developed a mathematical model of stratospheric chemistry based on reaction rates measured in the laboratory by chemical kinetics specialists. He considered all the reactions known at the time, and showed that the most effective mechanism for destroying ozone in the stratosphere is a catalytic cycle[9] with the catalyst element being nitric oxide (NO) even if it is present in small quantities in the atmosphere (Figure 6.8).

The question was immediately posed: Is nitric oxide really present in the atmosphere? And if so, in what proportion? And how is it formed? The answer was not obvious because, at this time, no measurements of nitrogen compounds had been made in the stratosphere. Nitric oxide was known to be produced above 80 km in altitude by the dissociation of molecular nitrogen (N_2) from solar radiation and energetic particles, but it was unlikely that nitrogen oxide could be transported downwards to reach the stratosphere before it is destroyed. In 1971, Crutzen and Nicolet independently proposed a mechanism by which nitric oxide would be formed in the stratosphere: it would result from the oxidation of nitrous oxide (N_2O), a gas that is produced by the action of different bacteria present in the soils. It is the "hilarious gas," well known especially by dentists who use it to relax their anxious patients. This molecule is sufficiently stable in the atmosphere to be transported to high altitudes and injected into the stratosphere. In other words, stratospheric ozone, which is produced by the action of ultraviolet solar radiation on oxygen, would be destroyed mainly by a process that results from biological processes in soils.

The first piece of observational evidence came from a closer inspection of infrared solar spectra recorded in 1968 using a balloon borne spectrometer floating at 30 km altitude and pointing toward the Sun tangentially to the Earth. David G. Murcray and his colleagues at the University of Denver detected absorption bands that they attributed to the presence of nitric

---

9. The reactions proposed by Crutzen to destroy ozone are $NO + O_3 \rightarrow NO_2 + O_2$ and $NO_2 + O \rightarrow NO + O_2$. All of these two reactions are equivalent to Chapman's $O_3 + O \rightarrow 2O_2$ process, but are considerably more efficient in the stratosphere.

Quart. J. R. Met. Soc. (1970), **96**, pp. 320-325

551.510.41 : 551.510.534

The influence of nitrogen oxides on the atmospheric ozone content

By P. J. CRUTZEN*

Clarendon Laboratory, Oxford University

(Manuscript received 5 November 1969, communicated by Dr. C. D. Walshaw)

SUMMARY

The probable importance of NO and NO_2 in controlling the ozone concentrations and production rates in the stratosphere is pointed out. Observations on and determinations of nitric acid concentrations in the stratosphere by Murcray, Kyle, Murcray and Williams (1968) and Rhine, Tubbs and Dudley Williams (1969) support the high NO and NO_2 concentrations indicated by Bates and Hays (1967).
Some processes which may lead to production of nitric acid are discussed.
The importance of O (^1S), possibly produced in the ozone photolysis below 2340 Å, on the ozone photochemistry is mentioned.

4. RESULTS

The daily production and destruction for odd oxygen below 50 km have been estimated for conditions at the Equator. The observed ozone distribution was taken from Dütsch (1964). The results of the computations are shown in Table 1.

It can be seen that the destruction of odd oxygen by the nitrogen oxides is of the same order of magnitude as the production by photodissociation of molecular oxygen. Reductions of odd oxygen by odd nitrogen is, according to these estimates, dominant between approximately 25 and 40 km. There are, however, indications in the Table that reduction by OH and HO_2 begins to be of larger importance around the stratopause.

6. CONCLUSIONS

There is a distinct possibility that nitrogen oxides are of great importance in ozone photochemistry. In the first place we urgently need observations on their concentrations in the stratosphere. Investigations about the photodissociation products of N_2O and its origin (see Bates and Hays 1967) should be continued and extended in order to establish if N_2O is an important source for odd nitrogen in the upper atmosphere. If, however, most of the stratospheric NO and NO_2 is produced at very high levels by other processes some solar cycle influence on the ozone layer will be possible.

Figure 6.8. The summary and some quotes of the paper published by Paul Crutzen on the influence of nitrogen oxides on stratospheric ozone, while he was a postdoctoral Fellow at Oxford University. Credit: Royal Meteorological Society; reproduced from *Quarterly Journal of the Royal Meteorological Society* 96 (1970), 320–25, 1970.

acid in the stratosphere (Figure 6.9). The Denver group deduced that nitric oxide should also be present in the lower stratosphere. However, the first quantitative in situ measurements of the nitric oxide (NO) concentration between 17 and 23 km altitude were made on March 16, 1973 in New Mexico by Brian Ridley and Harold Schiff (1923–2003), then at York University in

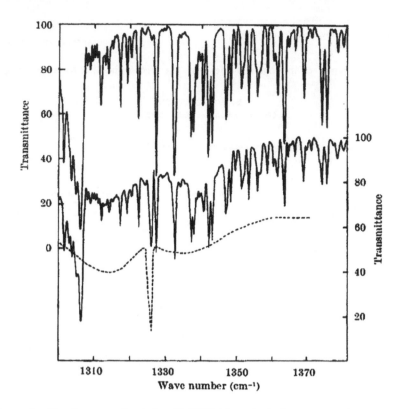

Figure 6.9. The discovery of a nitric acid (HNO₃) layer in the lower stratosphere. The solar infrared spectrum recorded on December 7, 1967, by the group of David G. Murcray at the University of Denver revealed the existence of a previously unidentified absorption feature at the wavenumber of 1,325.7 cm^{-1}. The Denver group concluded that this atmospheric absorption was due to the presence a nitric acid layer located between approximately 22 and 30 km altitude. The upper curve represents the solar spectrum observed by a balloon-borne spectrometer pointing to the Sun at 11 km altitude for a zenith angle of 58.7°. With the corresponding small absorption path, no significant spectral feature produced by HNO₃ could be detected. The lower curve represents the solar spectrum observed when the balloon was floating at 30 km and the spectrometer was pointing to the Sun at a zenith angle of 92.4°. This case is characterized by a longer optical path through the HNO₃ layer, so that the absorption feature due to nitric acid is clearly visible. The dashed line represents the location of the absorption line used to identify the presence of nitric acid. Weak absorption bands observed in the infrared solar spectrum, also subject to long path enhancement, were attributed by Murcray and his colleagues to the presence of nitrogen dioxide (NO₂) in the lower stratosphere. No estimate of the atmospheric abundance of HNO₃ and NO₂ were provided because no laboratory data about the strength of the corresponding absorption bands were available. Reproduced from Murcray et al., *Nature*, 218 (1968), 78–79.

Toronto, Canada, using a balloon-borne chemiluminescence instrument.[10] The Canadian team deduced from their observations that the mixing ratio of NO in the lower stratosphere is close to 0.1 ppbv with an uncertainty of 60%. Additional measurements were carried out subsequently up to the altitudes of 30 or 35 km; they confirmed the presence of nitrogen oxides (NO and NO_2) in most of the stratosphere, and thus reinforced, as Crutzen has highlighted, that ozone's primary destruction mechanism in the stratosphere is due to nitrogen oxides. The vulnerability of the ozone layer to nitrogen oxides that could be injected by humans into the stratosphere became a major concern (see Chapter 7).

Crutzen's discoveries and the resulting modifications to Chapman's original model made it possible to faithfully reproduce the vertical distribution of ozone in the upper stratosphere. However, there were still important discrepancies between the theory and the observation below 25 km, in an area of the atmosphere where ozone is mostly sensitive to the transport associated with atmospheric dynamics and meteorological processes.

In 1974, Crutzen moved to the United States where, in 1977, he became Director of the Air Quality Division at the National Center for Atmospheric Research (NCAR) in Boulder. He then became interested in other issues affecting humanity, particularly the global environmental impacts of the possible use of nuclear weapons. He contributed to the so-called "nuclear winter" project involving scientific authorities from the United States, Europe, and the Soviet Union, a project that demonstrated that a series of nuclear explosions in one part of the world would produce a thick layer of fine particles in the atmosphere all around the Earth. This would cause a considerable cooling of the entire planet's surface and would therefore affect the originator of the nuclear conflict as well as all other countries.

In 1980, Crutzen returned to Europe to become Director at the Max Planck Institute of Chemistry in Mainz, Germany. He succeeded Christian Junge (1912–1996), a pioneer of atmospheric aerosol research and air chemistry. During this period of his career, Crutzen evoked the possibility of counteracting the global warming produced by man-made CO_2 by injecting large quantities of sulfate particles into the stratosphere. These particles should scatter back to space a fraction of the incoming solar energy and

---

10. The principle of this technique is the measurement of light resulting from the chemiluminescent reaction $NO + O_3 \rightarrow NO_2 + O_2 + light$. The intensity of the light that is detected is a function of the atmospheric abundance of NO.

thus cool the Earth's surface. Crutzen's proposal generated a debate on the necessity, benefits, and dangers of climate engineering, and led to controversies in the scientific community and among economists, sociologists, and even philosophers. Crutzen did not advocate implementing this method to limit the extent of climate change, but believed that, if measures to reduce CO_2 emissions were not rapidly adopted, humanity would have no choice but to turn to this type of method. The effects and consequences must therefore be investigated.

Crutzen did not stop here. At a meeting of the International Geosphere Biosphere Programme (IGBP) held in February 2000 in Cuernavaca, Mexico, he suggested that the geological period in which we live is so dominated by the effects of human activity that it should be called *anthropocene*[11] rather than Holocene. His proposal opened a new debate, this time among geophysicists and geologists. In 1995, Crutzen was awarded the Nobel Prize for Chemistry for his work on atmospheric chemistry.

Ozone Destruction by Chlorinated and Brominated Compounds

At the beginning of the 1970s, the general feeling was that the ozone problem had been solved. In 1950, Chapman's theory had been corrected by Bates and Nicolet's scheme and 20 years later by Crutzen's theory. It was thought that the mechanisms for the formation and destruction of this molecule in the stratosphere were now fully understood. This feeling was supported by the fact that mathematical models that take into account these chemical mechanisms and simulate atmospheric transport produced a vertical distribution of ozone in accordance with the observation. It was without counting on a surprise that appeared in 1973.

At a scientific meeting held in Kyoto, Japan, in September 1973, two ionosphere specialists at the University of Michigan, Richard Stolarski and Ralph Cicerone (Figure 6.10), showed that chlorine is an additional element capable of catalyzing ozone depletion.[12] This idea could have gone rather

---

11. The word *Anthropocene*, introduced by Eugene F. Stoermer (1934–2012), a professor of biology at the University of Michigan, and popularized by Paul Crutzen, accounts for the major evolution in Earth's history generated by the direct intervention of human beings on nature.

12. The proposed reactions are $Cl + O_3 \rightarrow ClO + O_2$ and $ClO + O \rightarrow Cl + O_2$. The net effect of this catalytic cycle is equivalent to the loss proposed by Chapman: $O_3 + O \rightarrow 2O_2$.

Figure 6.10. Ralph Cicerone (left) and Richard Stolarski (right), who highlighted that chlorine atoms destroy ozone in the stratosphere. Credit American Geophysical Union (https://gec.agu.org/ralph-cicerone/): and https://www.youtube.com/watch?v=kEiGn3ooweg.

unnoticed because laboratory chemists knew for some time that chlorine destroys ozone, and because the presence of reactive chlorine had never been detected in the stratosphere. In fact, the study of the two Michigan researchers had been discreetly commissioned by NASA to estimate the possible effect on the ozone layer of the chlorinated chemical compounds emitted by the space shuttle. A confidential report dated June 3, 1973, and prepared by the Michigan team discussed the potential effects of the shuttle, but a related article on the effect of chlorine on ozone submitted to *Science* was rejected by the journal. The study was eventually published in the *Canadian Journal of Chemistry* (Figure 6.11). At the time of their investigation, Stolarski and Cicerone were not aware of the existence of a large anthropogenic source of chlorine in the atmosphere. Their study, however, was important to explain later on why the chlorofluorocarbons of industrial origin could deplete ozone in the stratosphere. In a subsequent study, Steven Wofsy of Harvard University added that bromine, also released to the atmosphere as industrial products, is an additional catalyst to destroy ozone.

In 1984, Stolarski left Michigan and joined NASA's Goddard Space Flight Center near Washington, District of Columbia, where he continued to work on ozone issues during the following decades. Cicerone (1943–2016) became Director of the Atmospheric Chemistry Division at the National Center for Atmospheric Research in Boulder where he succeeded Paul Crutzen. He then became professor and later chancellor of the University of California at Irvine before becoming the president of the US Academy of Sciences in 2005.

Stratospheric Chlorine: a Possible Sink for Ozone

R. S. STOLARSKI AND R. J. CICERONE

Space Physics Research Laboratory, The University of Michigan, Ann Arbor, Michigan 48105
Received January 18, 1974

This study proposes that the oxides of chlorine, ClO_x, may constitute an important sink for stratospheric ozone. A photochemical scheme is devised which includes two catalytic cycles through which ClO_x destroys odd oxygen. The individual ClX constituents (HCl, Cl, ClO, and OClO) perform analogously to the respective constituents (HNO_3, NO, NO_2, and NO_3) in the NO_x catalytic cycles, but the ozone destruction efficiency is higher for ClO_x. Our photochemical scheme predicts that ClO is the dominant chlorine constituent in the lower and middle stratosphere and HCl dominates in the upper stratosphere. Sample calculations are performed for several ClX altitude profiles: an assumed 1 p.p.b. volume mixing ratio, a ground level source, and direct injection by volcanic explosions. Finally we discuss certain limitations of the present model: uncertainty in stratospheric OH concentrations, the possibility that ClOO exists, the need to couple ClO_x cycles with NO_x and HO_x cycles, and possible heterogeneous reactions.

Figure 6.11. The abstract of the paper published in 1974 by Richard Stolarski and Ralph Cicerone indicating that the oxides of chlorine could provide an important loss mechanism for stratospheric ozone. Reproduced from Stolarski, R. S. and R. J. Cicerone, *Canadian Journal of Chemistry* 52 (1974), 1610–15.

The Role of Atmospheric Dynamics and Ozone Transport

In the early years of the twentieth century, not much information was available on the dynamics of the upper layers of the atmosphere. Meteorology was still in its infancy, but it was making substantial progress as a rigorous science. Under the influence of the "Bergen School" in Norway and in particular of prominent meteorologist Vilhelm Bjerknes (1862–1951), weather forecasts evolved from a relatively empirical exercise to a deterministic problem based on the Newton fundamental laws applied to the atmospheric fluid. Bjerknes's visionary statements published in 1904 paved the way to numerical weather predictions.[13] Following these concepts, British

---

13. In a paper entitled "Das Problem der Wettervorhersage, betrachtet vom Standpunkte der Mechanik und der Physik" (The problem of Weather prediction, as seen from the standpoints of mechanics and physics) published in January 1904 in the Meteorologische Zeitschrift, Vilhelm Bjerknes wrote, "If, as every scientifically inclined individual believes, atmospheric conditions develop according to natural laws from their precursors, it follows that the necessary and sufficient conditions for a rational solution of the problems of meteorological prediction are the following: 1: the condition of the atmosphere must be known at a specific time with sufficient accuracy; 2: the laws must be known, with sufficient accuracy, which determine the development of one weather condition from another." (Translation from the German.)

meteorologist Lewis Fry Richardson (1881–1953) made the first numerical forecast of the weather in Europe. While stationed in France as an
ambulance driver during the First World War, he attempted to solve by
hand the governing equations describing the behavior of the atmospheric
flow, but failed to reproduce the observed meteorological situation on the
European continent. After the war, the meteorology community realized
that progress required more cooperation and a sustained research effort.
Bjerknes, who was interested by the dynamical processes that drive the
general circulation of the atmosphere and in the methods of weather forecasting, introduced around 1920 the concept of the polar front, which plays
an important role in the evolution of weather patterns in the extratropical
regions. The International Commission for the Scientific Investigation of
Upper Air (Figure 6.12) provided an opportunity to enhance scientific
exchanges and, specifically, to further investigate the dynamical processes
at high altitude.

01 Martin KNUDSEN *(50)*
(Denmark; 1871-1949)
02 Axel WALLÉN *(44)*
(Sweden; 1877-1935)
03 Juan CRUZ CONDE *(xx)*
(Spain; 18xx-19xx)
04 Ernst CALWAGEN *(27)*
(Sweden; 1894-1925)
05 Oscar EDLUND *(29)*
(Sweden; 1892-1959)

06 Johan SANDSTRØM *(47)*
(Sweden; 1874-1947)
07 Theodor HESSELBERG *(36)*
(Norway; 1885-1966)
08 Willem van BEMMELEN *(53)*
(Netherlands; 1868-1941)
09 Vilhelm BJERKNES *(60)*
(Norway; 1861-1952)
10 Hilding KÖHLER *(33)*
(Sweden; 1888-1982)

11 Lewis Fry RICHARDSON *(40)*
(Great Britain; 1881-1953)
12 Paul Louis MERCANTON *(45)*
(Switzerland; 1876-1963)
13 Harald NORINDER *(33)*
(Sweden; 1888-1969)
14 Napier SHAW *(67)*
(Great Britain; 1854-1945)
15 Finn MALMGREN *(26)*
(Sweden; 1895-1928)

16 Jules JAUMOTTE *(34)*
(Belgium; 1887-1940)
17 Jacob BJERKNES *(24)*
(Norway; 1897-1975)
18 Ewoud van EVERDINGEN *(48)*
(Netherlands; 1873-1955)
19 Ernest GOLD *(40)*
(Great Britain; 1881-1976)
20 Sakuhei FUJIWARA *(37)*
(Japan; 1884-1950)

21 Alfred de QUERVAIN *(42)*
(Switzerland; 1879-1927)
22 Geoffrey I. TAYLOR *(35)*
(Great Britain; 1886-1975)
23 Philippe SCHERESCHEWSKY *(29)*
(France; 1892-1980)
24 Charles J.P. CAVE *(50)*
(Great Britain; 1871-1950)
25 Rikichi SEKIGUCHI *(35)*
(Japan; 1886-1951)
26 Gustav GYLLSTRÖM *(18)*
(Sweden; 1903-19xx)

Figure 6.12. Participants to the 8th meeting of the International Commission for the
Scientific Investigation of Upper Air chaired by Vilhelm Bjerknes in Bergen, Norway,
on July 25, 1921. The identification numbers run in five columns from back to front, each
followed by the person's given name, age at the time of the meeting in brackets, country
of work, and life span. Credit: University of Bergen Library, Norway.

Observations of ozone could provide interesting information about the meteorological situations and the general circulation of the atmosphere. The numerous observations made by Dobson and other investigators since the 1930s showed that the distribution of ozone in the atmosphere is very sensitive to synoptic-scale weather variations. From Herzberg's laboratory investigations, it was established that the formation of ozone by the photo-dissociation of molecular oxygen occurs only above 25 km altitude so that, below this altitude in the stratosphere, ozone must be a quasi-inert tracer that can be transported over long distances by the atmospheric circulation. Dobson also showed that ozone is considerably more abundant in the polar regions than above the equator, and that the concentration of this molecule reaches a maximum at the end of the winter. In 1929, he wrote,

> The only way in which we can reconcile the observed high ozone concentration in the Arctic in spring and the low concentration in the tropics [...] would be to suppose a general slow poleward drift in the highest atmosphere with a slow descent of air near the poles.

However, in the following years, Dobson questioned and then rejected his original idea because it seemed to violate the momentum conservation law.

During the Second World War, Dobson was invited to study the conditions conductive of the formation of condensation trails (contrails) produced by aircraft. This problem had become crucial for airmen who did not wish to be spotted by the enemy during bombing missions. Alan Brewer (1915–2007), a scientist at the British Meteorological Office, who at the beginning of the war had been assigned as shift weather forecaster for the Royal Air Force, was tasked to investigate with Dobson how aircraft could avoid the formation of these condensation trails. To address this question, the two scientists had to measure very accurately the temperature and humidity at the flight altitudes of these aircraft. Brewer designed new instruments to measure temperature and humidity that could be installed on high-altitude aircraft. The measurements of temperature and humidity in the upper troposphere were known to be very challenging and often very inaccurate. Brewer developed a frost point hygrometer,[14] and the first measurements of the humidity

---

14. Frost-point hygrometer: in this instrument, the temperature of a chilled mirror is controlled such that the mirror maintains a small and constant layer of frost coverage. The frost point temperature of the air passing over the mirror is equal to the measured temperature of the mirror. The frost point is the temperature at which

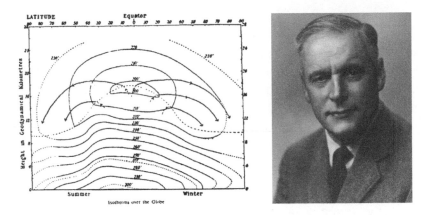

Figure 6.13. (Left panel) Conceptual representation of the meridional transport in the stratosphere known as the Brewer-Dobson circulation. Reproduced from Brewer, A. W., *Quart. J. Roy. Meteor. Soc.* 75 (1949), 351–63. (Right panel) Photograph of Alan Brewer. Reproduced from Bojkov (2012).

were carried out from a flying fortress (one of the six large aircraft given by Roosevelt to Churchill as part of the war effort). This aircraft had a ceiling altitude of 37,000 ft, and could occasionally penetrate in the stratosphere. Subsequent temperature and water vapor measurements were performed from a Mosquito aircraft that could reach higher stratospheric altitudes. These observations led to a major discovery: the stratosphere is considerably drier than the lower layers of the atmosphere. Brewer was himself surprised by his findings and wrote later in Year 2000[15]:

> I would not have believed before I had started that the air could be so dry, but I saw it with my own eyes. I had plenty of trouble with people convincing them that we could really measure that dryness [...]. The effects of the tropopause were very striking. Personally, I very soon formed the opinion that the very dry air had come from the equatorial tropopause.

This finding allowed Brewer (Figure 6.13) to establish a simple conceptual representation of the global mass circulation of tropospheric air through the

---

the air is saturated with respect to water vapor over an ice surface. Condensation trails are formed in the atmosphere when water vapor condenses and freezes around the small particles that are present in the exhaust of aircraft engines. They can be formed if the air temperature is lower than the frost point.

15. SPARC Newsletter number 15, July 2000.

stratosphere. In a seminal paper published in 1949 (Figure 6.14), Alan Brewer suggested the existence of a meridional stratospheric circulation in which tropospheric air enters the stratosphere in the tropics, moves from the tropics toward the poles, and descend at mid and high latitudes. This circulation was proposed to explain at the same time the dryness of the stratosphere that Brewer had observed (see Chapter 3) and the high concentrations of ozone that Dobson had reported in the polar regions (Figure 6.14). In his 1949 paper, he stated:

551.510.5

EVIDENCE FOR A WORLD CIRCULATION PROVIDED BY THE MEASUREMENTS OF HELIUM AND WATER VAPOUR DISTRIBUTION IN THE STRATOSPHERE

By A. W. BREWER, M.Sc., A.Inst.P.

(Manuscript received 23 February 1949)

SUMMARY

Information is now available regarding the vertical distribution of water vapour and helium in the lower stratosphere over southern England. The helium content of the air is found to be remarkably constant up to 20 km but the water content is found to fall very rapidly just above the tropopause, and in the lowest 1 km of the stratosphere the humidity mixing ratio falls through a ratio of 10—1.

The helium distribution is not compatible with the view of a quiescent stratosphere free from turbulence or vertical motions. The water-vapour distribution is incompatible with a turbulent stratosphere unless some dynamic process maintains the dryness of the stratosphere. In view of the large wind shear which is normally found just above the tropopause it is unlikely that this region is free from turbulence.

The observed distributions can be explained by the existence of a circulation in which air enters the stratosphere at the equator, where it is dried by condensation, travels in the stratosphere to temperate and polar regions, and sinks into the troposphere. The sinking, however, will warm the air unless it is being cooled by radiation and the idea of a stratosphere in radiative equilibrium must be abandoned. The cooling rate must lie between about 0·1 and 1·1°C per day but a value near 0·5°C per day seems most probable. At the equator the ascending air must be subject to heating by radiation.

The circulation is quite reasonable on energy considerations. It is consistent with the existence of lower temperatures in the equatorial stratosphere than in polar and temperate regions, and if the flow can carry ozone from the equator to the poles then it gives a reasonable explanation of the high ozone values observed at high latitudes. The dynamic consequences of the circulation are not considered. It should however be noted that there is considerable difficulty to account for the smallness of the westerly winds in the stratosphere, as the rotation of the earth should convert the slow poleward movement into strong westerly winds.

Figure 6.14. Summary of the paper published by Alan Brewer in 1949. Reproduced from Brewer, A.W., *Royal Meteorological Society 75* (1949), 351–63.

The observed distributions [of water vapor] can be explained by the existence of a circulation in which air enters the stratosphere at the equator where it is dried by condensation, travels in the stratosphere to temperate and polar regions, and sinks into the troposphere.

As mentioned above, Dobson was initially skeptical about this idea, but he eventually agreed with Brewer's views because it was consistent with the ozone transport that he had observed. The term "Brewer-Dobson circulation" is now universally used to describe a global-scale phenomenon that explains the transport of chemical species from the equator to the poles above the tropopause.

Brewer, who worked at Oxford University, but was born in Montreal where his parents lived for a short term, returned to Canada in 1962 to establish a Department of Meteorology at the University of Toronto. When he retired in 1976, he moved back to the United Kingdom to realize a long-term dream. He bought and operated a farm in Devon for almost 20 years. He died in Bristol in 2007.

Other meteorologists, particularly James R. Holton (1938–2004), professor at the University of Washington at Seattle and Michael E. McIntyre, mathematician at Cambridge University (Figure 6.15), further

Figure 6.15. Meteorologists James Holton (left) and Michael McIntyre (right), both pioneer researchers in upper atmosphere dynamics. Credit: https://atmospheres.agu.org/awards/holton-award/ and https://www.ae-info.org/ae/Member/McIntyre_Michael.

investigated the dynamical mechanisms that generate the Brewer-Dobson circulation and the associated large-scale stratospheric transport of chemical species. Their work showed that transport from the tropics to the polar regions results from the propagation and dissipation of large-scale "planetary waves," which are generated by the interaction of surface winds with large mountain ranges such as the Alps, the Himalayas, or the Rockies. These waves propagate upwards in the stratosphere, but, as shown by Swedish-born American meteorologist Carl-Gustaf Rossby (1896–1957), this propagation is only possible during winter when the prevailing winds are directed from west to east (westerly winds). As the amplitude of these winter waves increases with altitude, they become unstable and eventually break, just as ocean waves dissipate as they encounter the beaches. The energy and momentum brought by these waves to the upper atmosphere generate the meridional circulation proposed by Brewer and responsible for the transport of ozone from the tropical regions to the polar regions. This explains also why ozone accumulates at high latitudes during the winter before being photochemically destroyed in the spring when the Sun "returns" to the polar regions. In other words, the high levels of ozone observed in the 1930s by Dobson in the boreal regions are the result of large-scale dynamics, and the hemispheric asymmetry in the ozone distribution, also reported by Dobson, results directly from interhemispheric differences in the general circulation of the atmosphere. The orography of the southern hemisphere is indeed very different from that of the northern hemisphere, and the amplitude of planetary waves in the austral regions is usually considerably smaller than the amplitude observed in the boreal regions. In particular, a large vortex (called the *polar vortex*) develops throughout the winter around the Antarctic continent and, because it is very stable and persistent, constitutes a real barrier that prevents ozone from reaching the regions of the South Pole. Ozone therefore accumulates at the edge of the Antarctic region (60°S). In contrast to the situation observed in the Southern hemisphere, the polar vortex of the northern hemisphere, frequently disturbed by the presence of intense planetary waves, constitutes a relatively weak barrier against the poleward transport of ozone. This explains the large amounts of ozone present in the Arctic regions that Dobson had reported in the 1920s, and the comparatively smaller amounts observed in the Antarctic stratosphere during the southern winter.

Occasionally during winter, the amplitude of the planetary waves increases dramatically, especially in the Northern hemisphere, leading to strong disturbances in the zonal winds around the polar vortex, and producing dramatic increases in the polar stratospheric temperatures (about 50°C in some areas) in just a few days. A sudden stratospheric warming was observed for the first time on January 27, 1952, by German meteorologist Richard Scherhag (1907–1970), the founding Director of the Meteorological Institute at the Free University of Berlin who had directed the Upper Atmosphere Group at the "Reichswetterdienst" in Berlin before the Second World War. By analyzing radiosonde data, Scherhag noted an explosive increase in the stratospheric temperature over the city of Berlin, an event that was first called a "Berlin phenomenon" before it was established that the perturbation covers a large fraction of the Northern hemisphere. The group in Berlin continued to monitor and study these stratospheric disturbances, in particular under the leadership of meteorologist Karin Labitzke (1935–2015). Sudden stratospheric warmings greatly affect the meridional transport of ozone during several weeks and are believed to influence the surface weather.

Systematic observations of the structure of the atmosphere revealed the complexity of the stratospheric dynamics affected not only by large-scale planetary waves, but also by smaller gravity waves generated in the troposphere by orography and by weather disturbances. Diurnal and semi-diurnal tides forced by solar heating, lunar gravitational field, or latent heat release by convective systems also affect the dynamics of the upper atmosphere. A quasi-periodic reversal of the zonal wind with a period of approximately 28 months is also observed in the tropics. Richard Lindzen from the Massachusetts Institute of Technology (MIT) together with James Holton suggested in the 1970s that this "quasi-biennial oscillation" is driven by atmospheric waves emanating from the tropical troposphere. These complex and mutually interacting multi-scale waves (Figure 6.16) affect the distribution and transport of atmospheric chemical species and in particular of stratospheric ozone. General circulation models of the atmosphere, which account for the wave-driving of the Brewer-Dobson circulation, show that under increasing levels of atmospheric CO_2 and related climate change, the strength of the meridional circulation will increase in the future and hence affect the distribution of long-lived chemical species and the rate at which stratospheric ozone recovers of its anthropogenic depletion during the last decades.

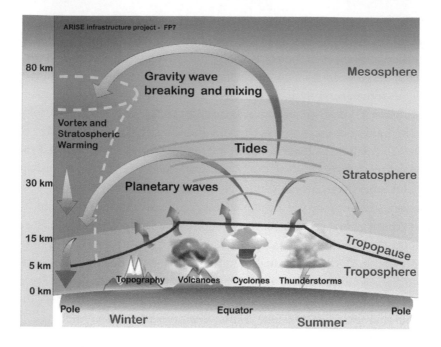

Figure 6.16. Schematic representation of dynamical processes affecting large-scale meridional transport in the different layers of the atmosphere. Upward propagating planetary waves (also called Rossby waves) are generated by the flow over the orography and by temperature contrasts between land and ocean. Gravity waves are produced by winds passing over mountain ranges and by weather disturbances including thunderstorms. Tides result from solar heating due to ultraviolet absorption by ozone and from latent heat release in the troposphere. These waves carry energy and momentum and interact with the mean zonal flow; their dissipation at high altitudes generates atmospheric mixing and produces the meridional circulation depicted by the arrows in the figure. Occasional wintertime amplification of the planetary waves leads to dramatic deviations of the atmospheric flow with a sudden warming of the stratosphere and the breakdown of the polar vortex. Credit: http://arise-project.eu/science.php.

Advanced Mathematical Models of Ozone

The mechanism introduced by Sydney Chapman in 1929 to describe the chemistry of ozone represented a hypothesis based on the understanding of photochemical processes at the time. The laws of chemical kinetics were applied to calculate the rates of production and destruction of ozone and atomic oxygen, and to infer the concentration of these two species in a stationary state, assuming that photochemical equilibrium conditions prevailed. Initially, Chapman suggested, on the basis of his calculations, that the maximum ozone concentration should be located near 50 km altitude.

Had Chapman known the correct values of the O_2 absorption coefficients in the ultraviolet, his estimates would have been closer to reality. Knowing from recently published observations that the ozone maximum was located around 22 to 25 km, Wulf and Deming in 1936 made some crude assumptions on the spectral distribution of the absorption coefficients to derive an ozone profile whose peak concentration is located near 25 km. The absolute values of the concentrations, however, were considerably overestimated, highlighting the fact that mathematical models do not provide a guaranteed element of certainty.[16]

It is therefore important to remember the role of models and their limitations. A model can be defined as a formal structure that is used to represent a set of phenomena between which certain relationships exist. It can be considered as a simplified representation of one or more processes and, in some cases, it allows one to visualize quantities that cannot be directly observed. Conceptually, models can be profoundly abstract and idealized, contributing, for example, to the development of new theories. If, on the other hand, they attempt to accurately represent the behavior of a system, they must deal realistically with the detailed and complex relationships that exist between the elements of this system. This is often accomplished by expressing the behavior of the system by a set of coupled differential equations and by solving these equations by numerical methods. Models make it possible to test hypotheses and, in particular, to highlight shortcomings of the theories. The inadequacy of the Chapman model led other researchers to complete the ozone mechanism proposed initially by adding the effects of destruction by compounds of hydrogen (Bates and Nicolet, 1950), nitrogen (Crutzen, 1970), and chlorine (Stolarski and Cicerone, 1974), and finally by considering the effects of atmospheric transport.

The good agreement between calculated and measured values led the scientists to believe that the ozone behavior was completely understood. However, as we will see later, there were surprises to come, in particular in the polar regions. As was the case, for example, with Newton's theory of gravity, first considered universal and definitive and then challenged by the theory of relativity, any "truth" of the moment can be called into question when new observations contradict the results derived from the theory.

---

16. This question is very well discussed in the small book by Jean Mawhin, *Les modèles mathématiques sont-ils des modèles à suivre?* (Should mathematical models be viewed as models to follow?) Royal Academy of Belgium, Collection L'Académie en Poche, 2017.

This was the case for the ozone issue; the theory had to be reviewed and new concepts had to be developed after the ozone hole was discovered.

In atmospheric chemistry problems, the development of mathematical models has often been constrained by the power of computers and, in particular, by their ability to process a large number of chemical species and to simulate the complexity of atmospheric dynamics and transport. The first models were therefore simple and one-dimensional; they provided only the mean vertical distribution of chemical species. Vertical exchanges were represented by a simple "eddy diffusion" formulation that has little in common with the complexity of dynamic exchanges. Tatsuo Shimazaki, a Japanese scientist working at the NOAA Aeronomy Laboratory in Boulder, Colorado, was among the first developers in the United States of such one-dimensional models. More detailed approaches accounted for the variations in physical and chemical quantities with latitude, but assumed uniformity of conditions with longitude. A pioneer in the development and implementation of such two-dimensional models was the Norwegian mathematician and meteorologist, Egil Hesstvedt (1920–1979), Professor at the University of Oslo (Figure 6.17). Hesstvedt had been working on the water budget in the upper atmosphere and specifically on the processes that lead to the formation of mother-of-pearl clouds (see Chapter 9 for more details). Several other two-dimensional chemical-dynamical models were established including, for example, the model developed jointly by Rolando Garcia (Figure 6.17) at the National Center for Atmospheric Research and Susan Solomon at the NOAA Aeronomy Laboratory, both at Boulder, Colorado. Finally, in an effort to realistically combine atmospheric dynamics and chemistry, three-dimensional chemical-dynamical models were initiated in the late 1960s and accounted for a substantial number of chemical reactions that occur between dozens of chemical species. The earliest three-dimensional ozone model was developed by the Australian scientist, Barrie G. Hunt, while working at the GFDL in Princeton. Hunt included in the GFDL atmospheric general circulation model two ozone-like "tracers" affected by photochemical processes, one in a "pure oxygen atmosphere" and the second one in an "oxygen-hydrogen atmosphere" (Figure 6.18). He was able to quantify the contribution of dynamical and photochemical chemical processes to the local rate of change in the ozone concentration at each point of the stratosphere. In the early 1970s, a similar global model was developed at the MIT under the leadership of Derek Cunnold (1940–2009; Figure 6.17). The advantage

Figure 6.17. Four pioneers in multi-dimensional chemical-transport modeling of the middle atmosphere. (Top panels from left to right) Egil Hesstvedt, professor at the University of Oslo, Norway (credit: https://lokalhistoriewiki.no/wiki/Eigil_Hesstvedt) and his successor Ivar Isaksen, professor in the same University. Credit: University of Oslo (https://www.mn.uio.no/geo/om/aktuelt/aktuelle-saker/2017/ivar-s-a-kristiansen-til-minne.html). (Lower panels from left to right) Rolando Garcia, senior scientist at the National Center for Atmospheric Research in Boulder, Colorado (credit: University Corporation for Atmospheric Research) and Derek Cunnold, research associate at MIT before becoming professor at the Georgia Institute of Technology in Atlanta, Georgia.

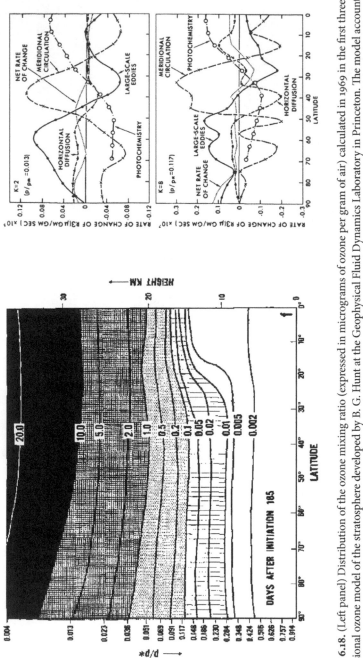

Figure 6.18. (Left panel) Distribution of the ozone mixing ratio (expressed in micrograms of ozone per gram of air) calculated in 1969 in the first three-dimensional ozone model of the stratosphere developed by B. G. Hunt at the Geophysical Fluid Dynamics Laboratory in Princeton. The model accounts for hydrogen chemistry in addition to the Chapman mechanism. The results shown here have been obtained by integrating the chemical transport equations over a period of 185 days. The three-dimensional model reproduces the increase of the mixing ratio with height as well as the low ozone abundance in the tropical lower stratosphere. (Right panel) Analysis of the different dynamical and chemical processes contributing to the calculated ozone tendency as a function of latitude at two different altitudes corresponding to the atmospheric pressure of 13 hPa or about 28 km (upper panel with $k = 2$) and of 117 kPa (lower panel with $k = 8$) or about 15 km, specifically the contribution of the transport by the meridional circulation and by planetary waves (large-scale eddies), horizontal diffusion, and the contribution of photochemistry (oxygen–hydrogen atmosphere). Reproduced from Hunt, B. G., Mon. Wea. Rev. 97 (1969), 297–306.

of general circulation models, when coupled to a chemical scheme, is that they allow one to quantify the respective influence of dynamical and chemical processes on the distribution of the ozone in the atmosphere. In all these chemical models, a system of partial differential equations is solved by advanced numerical methods with a spatial resolution that depends on the available computational resources. These modeling tools are used for interpreting atmospheric observations or predicting changes in the chemical composition of the atmosphere in response, for example, to human-induced disturbances.

The theoretical advances, and specifically the role of the catalytic cycles emphasized by Crutzen, Stolarski, and Cicerone, highlighted with increasing credibility the vulnerability of stratospheric ozone to human activities and specifically to the injections in the atmosphere of chemical species such as water vapor, nitrogen oxides, chlorine, and bromine. We will see in the following chapters that aviation, nuclear explosions, and the release into the atmosphere of long-lived bromine or chlorine containing products pose a threat to the ozone layer and thus to the global environment.

CHAPTER SEVEN

Ozone and Supersonic Aircraft

O
n March 2, 1969, the supersonic aircraft called *Concorde* and piloted by Major André Turcat (1921–2016), a graduate from Ecole Polytechnique in Paris, took off from Toulouse and made its first test flight in the skies of France. Europe wanted to be a forerunner in aeronautics and open up a new era for commercial aviation. The project, designed as part of an agreement between the French company *Sud-Aviation* (which later joined the consortium called *Aérospatiale*) and the *British Aircraft Corporation* (BAC), should have made Europe an industrial leader because it would have been able to compete with the activities of the American giants *Boeing* and *Lockheed*. Concorde entered service in 1976. It performed commercial flights over the Atlantic Ocean for a period of 27 years. Flights were permanently interrupted in 2003 three years after one aircraft of the fleet crashed after taking off from the Charles de Gaulle airport near Paris on its way to John F. Kennedy airport in New York. The Soviet Union developed a similar aircraft, the Tupolev 144, which was in service for only two years (1977 and 1978). The first test flights started in 1968 and were interrupted after the crash of a prototype brought to the 1973 Air show in Paris (Le Bourget airport). The airplane was nevertheless introduced in service on November 1, 1977, but the passenger fleet was permanently grounded after the crash of an improved model during a test flight in May 1978. It remained in service, however, as a cargo aircraft until 1983.

The American aeronautics industry that did not want to lose its technological and commercial lead, decided to accelerate the development of the Supersonic Transport, also called SST. This aircraft was expected to be larger and to fly faster (Mach 3 instead of Mach 2) and higher (20 km instead of 16 km) than Concorde. The project was entrusted to Boeing, which built two prototypes of the aircraft (the Boeing 2707-200) with the encouragement of the Federal Aviation Administration (FAA). Lockheed designed another model, essentially a larger Concorde.

The Effect of Water Released by Aircraft Engines

Despite the support provided by President Richard Nixon's administration, the SST project did not gain the unanimous support of the political class. Democratic Senator William Proxmire of Wisconsin (1916–2005), for example, considered the project to be a "frivolous" federal expenditure. The project was also severely criticized by environmental advocates not only because of the supersonic bang that would disturb the population living in the vicinity of aircraft wakes, but also because the water vapor released from aircraft engines into the stratosphere could alter, or even destroy, the ozone layer. As early as 1963, the problem was raised by chemist and meteorologist Jerome Pressman, who indicated that the water vapor released by future supersonic aircraft could have an adverse effect on the planet. Boeing became concerned by the growing public debate, and a scientist from the company's research laboratory, Halstead Harrison, using the model established by Barrie Hunt and later updated by Conway Leovy (1933–2011) at the University of Washington, calculated that water injected by a hypothetical fleet of 500 SST would reduce the thickness of the ozone layer by about 3.8%.

Another voice was heard in the late 1960s and early 1970s. James McDonald (1920–1971), a professor at the University of Arizona in Tucson (Figure 7.1), a member of the United States Academy of Sciences, and a proponent of the idea that unidentified flying objects—the famous UFOs—are of extraterrestrial origin, was also a fierce opponent of the supersonic aircraft against which he was conducting an intense campaign. During a hearing organized by an American parliamentary committee in 1970, he claimed that a fleet of 800 SST would cause 10,000 new cases of skin cancer in the United States. Massachusetts Republican Member of the House of Representatives, Silvio Ottavio Conte, who questioned McDonald and was a strong supporter of the SST project, claimed harshly that anyone "who believes in little green men" is not a credible witness for a parliamentary committee.

Figure 7.1. Harold Johnston (left) and James McDonald (right) who indicated that ozone could be destroyed by aircraft flying in the stratosphere. Reproduced from https://www2. lbl.gov/images/PID/Johnston.html and https://en.wikipedia.org/wiki/James_E._McDonald.

McDonald who heard laughs and giggles in the room felt publicly humiliated by the disparaging attitude of the member of the House and his colleagues. He recovered badly from this incident, which added to his personal problems, and after a first unsuccessful attempt, committed suicide in the Arizona desert on June 13, 1971.

The question of the possible pollution generated by the operations of a fleet of supersonic aircraft was discussed already in July 1970 at a workshop organized by MIT under the title "Study of Critical Environmental Problems (SCEP)." This meeting took place in preparation to the UN Conference on the Environment to be held in Stockholm in 1972. A working group chaired by meteorologist William Kellogg addressed different questions, including the impact of CO_2 increase on climate, the role of aerosols as cooling agents of the earth's surface, and the effect of a projected fleet of supersonic aircraft on climate and on the ozone layer. It concluded that

> no problem should arise from the introduction of carbon dioxide, and the reduction of ozone due to the interaction by water vapor or other exhaust gases should be insignificant […] Both carbon monoxide and nitrogen oxides can also play a role in stratospheric photochemistry, but these contaminants would be much less significant than the added water, and may be neglected.

The group attending the meeting, however, expresses some doubts and some concern:

> A feeling of genuine concern has emerged from these conclusions. The projected SSTs can have a clearly measurable effect in a large region of the world and quite possibly on a global scale.

The Department of Transportation (DOT) became increasingly concerned by the growing opposition against the SST and established at the end of 1970 a "Task Force" to examine all the problems affecting the development of the aircraft. DOT decided to bring together key scientists who had expressed views on the subject and asked them to formulate a position that represented a consensus. A meeting of the "Chemistry Subgroup of the Department of Commerce Panel on Supersonic Transport Environmental Research" was held at the National Center for Atmospheric Research (NCAR) in Boulder from March 18–19, 1971, and was chaired by Joseph O. Hirschfelder (1911–1990), a Professor at the University of Wisconsin. Hirschfelder was a renowned physicist who had participated in the Manhattan Project during the Second World War and had contributed to the development of the atomic bomb at the Los Alamos Laboratory. Among the participants were meteorologist William Kellogg (1917–2007), deputy director of NCAR; James McDonald (1920–1971); Julius London (1917–2009), professor of Meteorology at the University of Colorado (Figure 7.2); Frederick Kaufman (1919–1985), professor of Chemical Kinetics at the University of Pittsburg (Figure 7.2); Harold (Hal) Johnston (1920–2012), professor of chemistry at the University of California at Berkeley (Figure 7.1); Edwin F. Danielson (1921–1994), scientist at NCAR; as well as Arnold Goldburg, "Chief Scientist" of the SST Division at Boeing. The debates were tense because the protagonists' points of view were often incompatible. Kellogg, the chairperson of the first morning session on March 18, expressed the view that there is little to worry about: the changes in the ozone layer produced by an SST fleet should be much smaller than the magnitude of the natural variations occurring daily in the atmosphere. London presented a mathematical model that he had developed with his student Jae Park. Calculations showed that the water vapor emitted by a fleet of 500 SSTs would cause an ozone column reduction of only 1.2%. As for the effect of nitrogen oxides emitted by aircraft and hitherto ignored in the public debate, they would reduce the ozone column by 1.8%. These disturbances were considered by London to be relatively small and, based on this model, some participants in the meeting concluded that there is no cause

Figure 7.2. (Left) Chemist Frederick Kaufman (University of Pittsburg) and (right) meteorologist Julius London (University of Colorado) who attended the Boulder meeting on March 18–19, 1971, and expressed different points of views about the potential impact of a fleet of supersonic aircraft on the ozone layer. Source: National Academy of Sciences and http://www.irc-iamas.org/resources/jlondon/jlondon.php.

for alarm. Other presentations were made during that same day including a lecture by Danielsen on stratospheric transport and a paper by McDonald on the epidemiology of skin cancer. McDonald's presentation was "viciously attacked and interrupted in the middle of his talk over and over again" (to quote Harold Johnston[1]) by the representative of Boeing, Arnold Goldberg. However, at the end of the first day, the feeling resulting from the presentations and the discussions was that the impact of the SST on the ozone layer would be sufficiently small to be ignored. At the close of the session, an important news arrived from Washington and was communicated by the chair to the audience: The House of Representatives had decided by 215 against 204 votes to cut federal funding for the development of the SST.[2] This decision was motivated by economic considerations and perhaps related to the

---

1. Interview of Harold Johnston conducted by Sally Smith Hughes, PhD, at Berkeley, California, in 1999.

2. On March 25, 1971 the US Senate confirmed this decision by 51 against 46 votes. In May 1971, an attempt was made by a few senators to revive the SST project, but on May 19, the Senate turned down funds for the SST by a vote of 58 to 37.

inconvenience of the sonic boom. The threat of supersonic aviation for the ozone layer was not known by Congress at that time. The story seemed to be over. Nevertheless, the group decided to continue its discussion on the next day.

Already during the first day of the Boulder meeting, the chemists expressed their strong disagreement with the view that the impact of the SST would be negligible, and the debate intensified during the second day. Kaufman, for example, pointed out that there were very large uncertainties in the values of chemical reaction rates adopted in the London and Park model. Some of these coefficients had not even been measured in the laboratory, so that the values produced by the model were certainly incorrect and, in the best case, inaccurate. Kaufman referred to the value of a specific reaction rate constant[3] adopted by London and Park. He claimed that attempts had been made in his laboratory to measure this rate constant in 1964. The value was so small that, at best, they could provide only an upper limit.[4] Harold Johnston supported this point of view and noted that the model contained only four reactions to describe the nitrogen oxide chemistry. He stated a few days after the meeting that one of the kinetic coefficients adopted for the calculations[5] was overestimated by a factor of 13,000. While attending the meeting in Boulder, Johnston spent a large fraction of his time to read London's paper. He could not understand why the model that did not include the smog reactions[6] was exhibiting an ozone increase in the lower stratosphere in response to the aircraft NOx emissions. He became convinced that there was a mistake in the calculations. During night after the first day's meeting, Johnston wrote a 16-page document presenting his views and discussing how catalytical cycles involving nitrogen oxides effectively destroy ozone in the stratosphere. He made the point that the ozone loss due to the nitrogen

---

3. Reaction rate constants determine the rate at which chemical reaction occur.

4. Harold Johnston quotes Fred Kaufman during the Boulder meeting: "I appreciate what you guys are trying to do. People want you to make these calculations, and I sympathize with what you are up against. But, this is not what I said in 1964. I did not say the reaction rate is equal to 5×10^{-13}; I said that it was less than 5×10^{-13}. If it had been equal to 5×10^{-13} or faster, we could have measured it in our apparatus. We couldn't measure a thing. If you want to assume it is 5×10^{-13}, go ahead and assume it, but don't put my name down as justifying it."

5. The reaction involved was $O_2 + 2NO \rightarrow 2NO_2$ whose adopted rate was 1.0×10^{-33} cm^3/s rather than the correct value 7.6×10^{-38} cm^3/s.

6. Reactions that lead to the formation of tropospheric ozone in the presence of nitrogen oxide and organic compounds (see Chapter 10 for more details).

oxides present in the SST exhaust could exceed, perhaps by a factor 10, the ozone destruction attributed to water vapor. This document was distributed to the panel the next morning. Many of its members, however, remained unconvinced and claimed that Johnston was ignoring that the dispersion of the aircraft effluents by the air motions would considerably reduce the impact of aviation nitrogen oxides. A motion introduced by Johnston and stating that the oxides of nitrogen represent an important variable in the problem was tabled by an overwhelming majority at the end of the meeting. Interestingly, Crutzen, who had published his seminal paper on the effect of NOx on stratospheric ozone, had not been invited to the Boulder meeting and his work was largely ignored by the participants in the meeting.

The Effect of Nitrogen Oxides Released by Aircraft Engines
The Boulder discussion ended without consensus. In the following months, and despite the fact that the federal funding for the development of the SST had been cut, Congress established a research program that would investigate within four years the poorly understood dynamical and chemical processes of the stratosphere and predict the impact of a hypothetical fleet of 500 SSTs on the ozone layer. The organization of such a crash program became particularly urgent after Harold Johnston published on August 6, 1971, a resounding article in the scientific journal *Science* (Figure 7.3).[7] Johnston, who had just learned about Crutzen's work in Europe, stated in his paper that

> Oxides of nitrogen from SST exhaust pose a much greater threat to the ozone shield than does an increase in water.

He estimated that nitrogen oxides could cause a reduction in the ozone column not by a few percent, as suggested by London and Park, but by as large as a factor of two. This would allow solar radiation to penetrate very deeply into the lower atmosphere and cause adverse health effects, particularly a large number of skin cancers. In his paper, Johnston states

---

7. A first version of the paper that did not refer to Crutzen's (1970) paper was submitted to *Science* in May 1971. Johnston received the comments made by the reviewers together with a note from the Editor of the magazine requiring that Crutzen's work be acknowledged in the paper. This came as a surprise to Johnston who was not aware of Crutzen's study and who believed at this moment that he had made the discovery now attributed to Crutzen.

Reduction of Stratospheric Ozone by Nitrogen Oxide Catalysts from Supersonic Transport Exhaust

Abstract. *Although a great deal of attention has been given to the role of water vapor from supersonic transport (SST) exhaust in the stratosphere, oxides of nitrogen from SST exhaust pose a much greater threat to the ozone shield than does an increase in water. The projected increase in stratospheric oxides of nitrogen could reduce the ozone shield by about a factor of 2, thus permitting the harsh radiation below 300 nanometers to permeate the lower atmosphere.*

Figure 7.3. Abstract of Harold Johnston's paper published by *Science* magazine. Reproduced from Johnston, H. S., *Science* 173 (1971), 517–22.

in the "References and Notes" that his report in *Science* is an outgrowth of the discussion that took place at the March 18–19, 1971, meeting in Boulder, and he thanked the attendees at the meeting "for advice and for constructive opposition." More detailed conclusions (see Box 7.1) appeared in a 70-page internal laboratory report prepared by Johnston at Berkeley.

Johnston's study was speculative. And the question was immediately posed: Is there any evidence that nitrogen oxides injected into the stratosphere destroy the ozone layer? A full-scale test that affected the atmosphere provided at least a partial answer: Henri M. Foley (1917–1982) and Melvin A. Ruderman, two scientists from Columbia University in New York, claimed in 1973 that the Soviet-American nuclear explosions of the late 1950s and early 1960s produced amounts of nitrogen oxides comparable to the amount that would be injected each year into the stratosphere by a fleet of 500 supersonic aircraft. If the claim made by Johnston was correct, thermonuclear explosions in the atmosphere should have depleted considerable amounts of stratospheric ozone. The heat released by such powerful explosions is known to dissociate the nitrogen molecules (N_2) present in the atmosphere, and the nitrogen atoms that are produced combine with atmospheric oxygen (O_2) to produce nitrogen oxide (NO). However, no significant ozone depletion was observed in the months following the explosions, although this result was questioned by several scientists. Thus, at first, the impact of supersonic aviation appeared to be tiny, but the exact magnitude remained uncertain.

The research program led by the US Department of Transportation, known as the Climatic Impact Assessment Program (CIAP), offered scientists the opportunity to investigate the processes that affect ozone above the tropopause for the first time in a very comprehensive way. Europe, however,

Box 7.1. The ten conclusions of 1971 Johnston's study

1. If NO and NO_2, as such, build up in the stratosphere to the expected concentrations from SST operation, the ozone shield would be reduced by a large amount, about a factor of two.

2. If NO and NO_2 are converted to HNO_3 (or other inert molecules) at rate faster than is indicated by present knowledge, then the effect of NO from SST exhaust would be less than expected in (1) above. The chemistry of the stratosphere is sufficiently complicated that one should look for new, unexpected chemical reactions.

3. A large reduction in stratospheric ozone would be expected to change the temperature, structure, and dynamics of the stratosphere, which may modify the quantitative aspects of conclusion (1). These effects are outside the scope of this report.

4. In the reduction of ozone, the oxides of nitrogen at low concentrations exhibit a threshold effect, and at high concentrations the oxides of nitrogen reduce ozone according to the square root of NOx.

5. Transport by air motions has a major effect in shaping the vertical profile of ozone in the lower half of the stratosphere; the catalytic chemical action of NOx in destroying ozone is a large effect under all conditions of temperature, pressure, composition, and radiation distribution in the lower half of the stratosphere. Transport by air motions is an action that moves the large catalytic effect of NO from one part of the world to another, but it does not cancel this strong catalytic effect.

6. In the present stratosphere, oxides of nitrogen act to limit the concentration and partly to shape the distribution of ozone; the indicated mole fraction of NOx is about 10^{-9} at 20 km increasing to substantially higher values at higher elevations.

7. At all levels of the stratosphere, water vapor has less effect on ozone than the effect of natural NO on ozone. The most important effect on ozone by water in the stratosphere is its role in converting NO_2 to HNO_3, not the role of its free radicals (HOx) in directly destroying ozone. NOx from SST is a much more serious threat than water vapor with respect to reducing the ozone shield.

8. Further experimental studies, especially chemical analyses in the stratosphere and photochemical and kinetic studies in the laboratory are needed to clear up the uncertainties under (2) above.

9. Quantitative physiological studies should be made of what would happen to plants, animals, and people if the ozone shield should be reduced by various amounts up to a full tenfold reduction.

10. Even though it may be too complicated for a complete theoretical treatment, the stratosphere is vulnerable to added oxides of nitrogen, and forethought should be given to this hazard before the stratosphere is subjected to heavy use.

was not convinced that the American authorities were unbiased toward European aeronautical programs and therefore decided to organize its own research on the topic. In 1972, the French government established a Committee called "Comité pour l'étude des conséquences des vols stratosphériques" (COVOS) (committee to study the consequences of stratospheric flights) chaired by Professor Edmond Brun (1898–1979), a specialist of thermodynamics, a former student of Charles Fabry in Marseilles, and a member of the French Academy of Sciences. The United Kingdom set up a similar committee, the "Committee on Meteorological Effects of Stratospheric Aircraft" (COMESA), which worked under the leadership of meteorologist Robert J. Murgatroyd (1916–2010). In Australia, the Academy of Sciences also decided to perform a study on the subject. Measurement campaigns were organized by the various programs to probe the stratosphere, and mathematical models were established to accurately estimate the impact of a future fleet of supersonic aircraft.

In 1972, a first report on the subject was published by the Australian Academy of Sciences. It stated that

> Although there has already been a considerable amount of lower stratosphere flying over the last decade, the ozone concentrations have not decreased [...] We therefore believe that the effect of supersonic aircraft on the ozone layer is not likely to be serious ...

The conclusions of the CIAP program presented in 1975 were less optimistic; this report stated, for example, that the operation of 40 Concorde aircraft (flying between 15 and 17 km altitude) and of 820 Boeing aircraft (flying between 18 and 21 km) would lead to a decrease in the column ozone of the order of 13 to 17%. It concluded that the commissioning of 500 Concorde aircraft flying at an altitude of 16 km would increase the number of skin cancers in the United States by 20,000 per year, much more than James McDonald had envisioned a few years earlier. The DOT's report added that nitrogen oxides emitted by the US supersonic aircraft flying at 20 km altitude would destroy 4.5 times more ozone than an equal number of the Concorde. The French and English reports were less alarmist: they concluded that the number of supersonic aircraft that should be built in Europe was insufficient to significantly disrupt the ozone layer.

The presentation of the CIAP conclusions by the officials of the Program was strongly criticized by several members of the scientific community who accused the DOT to misrepresent the scientific findings (see Box 7.2) and

Box 7.2. The ozone war

In 1970, the controversy over the development of the civil supersonic aircraft (also called the SST) was in full swing in the United States. Several scientists as well as environmental groups were highlighting the risks: this airplane would not only generate a sonic boom, but its emissions at high altitude would be detrimental to the ozone layer. The problem quickly became political, and soon the US Senate got involved. Democratic Senators Birch Bayh of Indiana and Frank Church of Idaho proposed to adopt a so-called Stratospheric Protection Act of 1971. Washington State Senator Henry Jackson (1912–1983), often called by his opponents "The Senator from Boeing," tried to rescue the SST project by proposing to establish a research program on the stratosphere, which hopefully would definitively demonstrate that the fears expressed by some scientists were unfounded. The law proposed by Jackson was adopted in September 1971 and the project, named Climatic Impact Assessment Program (CIAP), was immediately initiated by the DOT under the supervision of Alan Grobecker (1915–1998), who had been working at the Institute of Defense Analysis. The objective of CIAP was "to assess by a report in 1974 the impact of climatic changes which may occur from the operation of a worldwide stratospheric fleet in 1990 in order to determine the regulatory constraints on flight in the stratosphere required to prevent adverse environmental effects."

Scientists were first delighted with the introduction of this program, which would allow them for the first time to thoroughly investigate the chemical and dynamic processes in this atmospheric layer that was full of poorly understood processes, and was therefore often called "the ignorosphere." The budget allocated for this four-year study was of the order of $23 million, which enabled researchers to set up new measurement projects and to develop advanced mathematical models of the atmosphere. Each year, progress reports were presented to several hundred researchers during a large international symposium organized at the DOT facility in Cambridge, Massachusetts, near the MIT campus.

The views expressed by scientists at these meetings were often not unanimous, and generated intense debates. Meteorologists often considered that the destruction of ozone would not be perceptible because the amplitude of the changes should be much smaller than the amplitude of the natural variations produced by the continuous dynamical disturbances of the atmosphere. The chemists including Harold Johnston pointed out that laboratory studies unequivocally show that ozone is being rapidly destroyed by nitrogen oxides and therefore supersonic aircraft flying at high altitude should deplete considerable amounts of ozone. Ozone depletion should result in a significant increase in the number of skin cancers. The tension between participants gradually amplified and became what Lydia Dotto and Harold Schiff call in their 1978 book an *ozone war*. Johnston's point of view was strongly disputed by some of his colleagues who claimed that no significant ozone depletion had been observed in the months following the thermonuclear explosions even though these explosions should have produced large quantities of nitrogen oxides in the atmosphere. Other scientists claimed that stratospheric motions would disperse the gas released by aircraft engines before they could act on atmospheric ozone. Others were wondering why to be alarmed by an increase in the intensity of the solar ultraviolet when we know that it corresponds to the increase in the dose received by a resident who would have moved South over the distance from New York to Washington, District of Columbia.

The results of the CIAP study were published in 1974 as six large volumes (a total of 9,000 pages; Figure 7.4) written by the group of scientists involved in the program. As with any such report, the scientific results were summarized in a "Report of Findings," written primarily for the country's political and economic authorities. Scientists were suspicious of a DOT that clearly supported the development of supersonic aircraft, and the report written by the Program Director Alan Grobecker, his Deputy Samuel Coroniti and their supervisor, Robert Cannon, all working for the federal government, provoked a new controversy. The document appeared to be much more reassuring than suggested by the scientific investigation. It stated, for example, that nitrogen oxide emissions from aircraft engines were expected to decrease significantly in the coming years due to future developments in combustion technology; therefore, the impact of aircraft should be "in the noise."

The reaction of scientists was immediate and was exacerbated by the content of a reassuring press conference organized in early 1975 by CIAP to summarize the outcome of the four-year investigation. The next day, the title of the Washington Post was clear, but inconsistent with the views of the scientists: "SST cleared on ozone." Under the headline "SST's No Peril Yet, Study says," the San Francisco Chronicle wrote in its edition of January 22, 1975: "A three-year study concludes that supersonic aircraft now flying will not damage the earth's protective blanket of ozone, the Department of Transportation said yesterday. Alan J. Grobecker, who directed the study, said a U.S. fleet of high-flying planes would not have weaken the ozone shield either."

In a rebuttal article published by the journal *Science*, Thomas Donahue, Chairman of the Department of Atmospheric and Oceanic Sciences at the University of Michigan, accused Alan Grobecker of denying the results of the study, and to discredit scientists like McDonald, Crutzen, and Johnston, who had raised the alarm and were now accused of providing damaging counsel to the country.

The CIAP program of the Department of Transportation ended in early 1975, and much of the ozone research was transferred to NASA, which developed an "Upper Atmosphere Research Program" (UARP). A series of satellites were launched to study the behavior of the ozone layer and to detect a possible long-term decrease in ozone concentration. At that point in time, an armistice between the protagonists of the "ozone war" was concluded, at least temporarily.

Figure 7.4. Harold Johnston at the University of California Berkeley behind stacked documents produced to assess the impact of supersonic aviation on the ozone layer and on climate. Reproduced from Johnston, H., *Annual Rev. Phys. Chem.* (1992).

to claim that the SST would not substantially disrupt the ozone layer. A few months later, however, a new piece of information, questioned some of the most alarmist points of view. New measurements of a reaction rate constant carried out by Carl Howard at the NOAA Aeronomy Laboratory in Boulder, showed that the rate constant adopted by CIAP for the key reaction[8] that leads to ozone formation in the lower atmosphere had been considerably underestimated. When corrected in the models, the predictions of ozone destruction by supersonic aircraft were substantially reduced. It even appeared that subsonic commercial aircraft flying at an altitude of 10 or 12 km and whose jet engines also eject nitrogen oxides could produce—rather than destroy—ozone in the upper troposphere. This latest information provided the scientific basis for the US DOT in 1976 to authorize the Concorde to land in New York and in Washington, District of Columbia. As for the American SST, it was not be built because its economic profitability was not established and its impact on the environment could not be ignored. Each of the two SST prototypes developed by Boeing, which never took off, was exhibited in an aviation museum.

---

8. $HO_2 + NO \rightarrow OH + NO_2$.

CHAPTER EIGHT

Ozone and Chlorofluorocarbons

The communication presented by Richard Stolarski and Ralph Cicerone in September 1973 at the Kyoto meeting (see Chapter 6), indicating that chlorine is an ozone-depleting agent, was not immediately perceived as particularly important because the presence of chlorine in the stratosphere had never been established. Chemists had long observed in their laboratories that chlorine reacts with ozone, but the issue was considered to be purely academic. If chlorine was present in the stratosphere, even in small quantities, one had to establish its origin. Stolarski and Cicerone in their presentation of 1974 evoked some possible natural sources: volcanic eruptions and ocean emissions. But what they did not say, because they weren't allowed to do so, was that their study was commissioned by NASA to avoid a campaign against the space shuttle similar to what happened to the American supersonic transport. The Shuttle, the flagship of American space exploration, released chlorine, and this chlorine could potentially produce small holes in the ozone layer as the spacecraft passes through the atmosphere after take-off. NASA therefore hoped that outside expertise from two independent scientists could reassure those concerned about the impact on the environment of rocket and space shuttle launches. In fact, public opinion never took on this problem.

Ozone, Chlorine, Bromine, and the Cold War

In the 1960s, during the Cold War, the use of chlorine and even bromine to catalyze ozone depletion was already on the agenda in American government and military circles. At the time, the idea of controlling weather and even climate by human intervention was considered a respectable idea. Several projects in this direction were therefore being developed. One of them, proposed by meteorologist Harry Wexler (1911–1962), the initiator of atmospheric observation satellites, was to use ballistic missiles to inject chemical compounds into the upper atmosphere. In December 1961, Wexler consulted chemist and spectroscopist Oliver R. Wulf at the California Institute of Technology in Pasadena on the possibility of producing a hole in the ozone layer over the Soviet Union. Such action would have enhanced the incident ultraviolet radiation reaching the surface in this part of the world and would have significantly reduced the rival nation's agricultural production with adverse impacts on the population. Three months later, Wulf recommended injecting chlorine or bromine at high latitudes, and Wexler concluded that injecting 100,000 tons of bromine into the stratosphere "should prevent ozone from forming at latitudes above 65 degrees N" (Figure 8.1). Calculations carried out on this occasion showed that at 35° of latitude, the corresponding decrease in ozone in the lower stratosphere

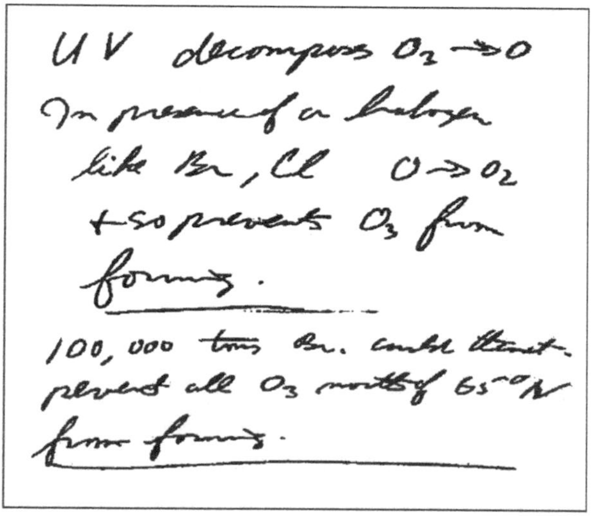
UV decomposes $O_2 \rightarrow O$
In presence of a halogen
like Br, Cl $O \rightarrow O_2$
+ so prevents O_3 from
forming.

100,000 tons Br. could then
prevent all O_3 north of 65° N
from forming.

Figure 8.1. A handwritten note by Harry Wexler in 1962 indicating that the injection of 100,000 tons of bromine would prevent ozone from forming at latitudes above 65° North. Document provided by James Rodger, Colby College.

would lead to a cooling of 10°C, 45°C, and 80°C at the respective altitudes of 12, 20, and 32 km. What the report did not clearly indicate was that the effect would not be limited to the Soviet Union alone, but that Europe and the United States would also be affected by this unilateral action. Fortunately, the concept was never implemented.

Wexler was one of the most influential meteorologists in the United States during the 1950s and early 1960s. He graduated in mathematics from Harvard in 1932 and was awarded his PhD in meteorology at MIT under the mentorship of Carl-Gustaf Rossby (1898–1957) and Bernhard Haurwitz (1905–1986). After 1934, he worked at the US Weather Bureau and during the Second World War worked as a captain and later a major and a lieutenant colonel with the weather service of the Army Air Corps. In 1944, he became the first scientist to fly into a hurricane to collect scientific data. The airplane was a Douglas A-20 Havoc. After the war, he returned to the US Weather Service where his influence steadily increased. He expressed interest in the study of the atmosphere of planets other than the earth, and became a pro-motor of the use of satellites in meteorology. In 1961 he served as the chief negotiator with the Soviet Union for the joint use of weather satellites. He liked to lecture on inadvertent and purposeful human-generated changes in large-scale phenomena in the atmosphere and climate engineering through manipulations of the Earth's radiative budget and of the stratospheric ozone layer. His life was abruptly interrupted during a vacation at Woods Hole in August 1962. And as scientific historian James R. Fleming writes,

> The idea that chlorine and other halogens could destroy stratospheric ozone was published in 1974,[1] while CFC production expanded rapidly and dramatically between 1962 and its peak in 1974. Had Wexler lived to published his ideas, they would certainly have been noticed and could have led to a different outcome and perhaps an earlier coordinated response to the issue of stratospheric ozone depletion.

Effect of Chlorofluorocarbons

In 1974, an article published in the journal *Nature* (Figure 8.2) by Mario Molina and F. Sherwood Rowland (1927–2012) of the University of California at Irvine provided the answer to the question raised by the study by Stolarski and Cicerone: is there an important source of chlorine in the stratosphere?

---

1. By Stolarski and Cicerone. See Chapter 6.

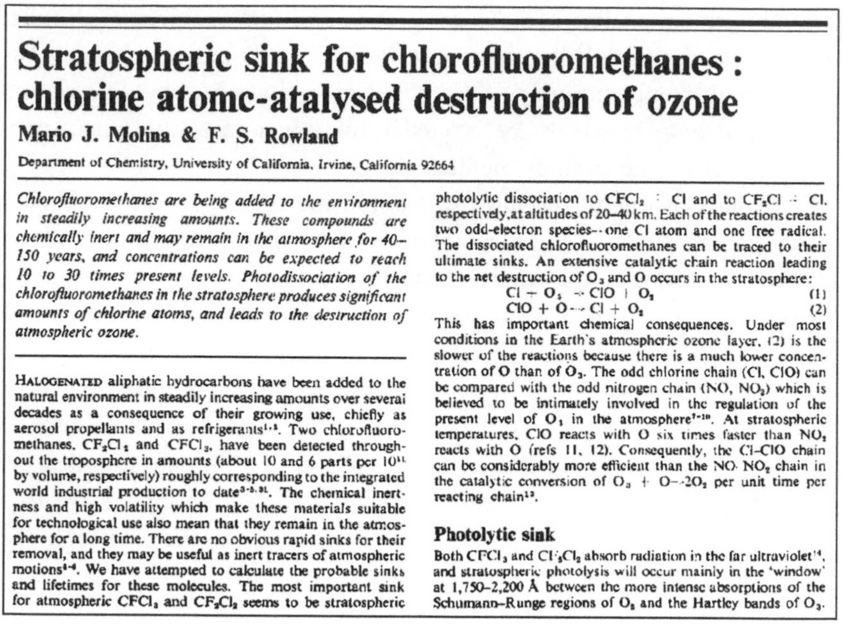

Stratospheric sink for chlorofluoromethanes : chlorine atomc-atalysed destruction of ozone

Mario J. Molina & F. S. Rowland

Department of Chemistry, University of California, Irvine, California 92664

Chlorofluoromethanes are being added to the environment in steadily increasing amounts. These compounds are chemically inert and may remain in the atmosphere for 40–150 years, and concentrations can be expected to reach 10 to 30 times present levels. Photodissociation of the chlorofluoromethanes in the stratosphere produces significant amounts of chlorine atoms, and leads to the destruction of atmospheric ozone.

HALOGENATED aliphatic hydrocarbons have been added to the natural environment in steadily increasing amounts over several decades as a consequence of their growing use, chiefly as aerosol propellants and as refrigerants[1-3]. Two chlorofluoromethanes, CF_2Cl_2 and $CFCl_3$, have been detected throughout the troposphere in amounts (about 10 and 6 parts per 10^{11} by volume, respectively) roughly corresponding to the integrated world industrial production to date[3-5,21]. The chemical inertness and high volatility which make these materials suitable for technological use also mean that they remain in the atmosphere for a long time. There are no obvious rapid sinks for their removal, and they may be useful as inert tracers of atmospheric motions[4-6]. We have attempted to calculate the probable sinks and lifetimes for these molecules. The most important sink for atmospheric $CFCl_3$ and CF_2Cl_2 seems to be stratospheric photolytic dissociation to $CFCl_2$: Cl and to CF_2Cl :: Cl, respectively, at altitudes of 20–40 km. Each of the reactions creates two odd-electron species--one Cl atom and one free radical. The dissociated chlorofluoromethanes can be traced to their ultimate sinks. An extensive catalytic chain reaction leading to the net destruction of O_3 and O occurs in the stratosphere:

$$Cl + O_3 \rightarrow ClO + O_2 \qquad (1)$$
$$ClO + O \rightarrow Cl + O_2 \qquad (2)$$

This has important chemical consequences. Under most conditions in the Earth's atmospheric ozone layer, (2) is the slower of the reactions because there is a much lower concentration of O than of O_3. The odd chlorine chain (Cl, ClO) can be compared with the odd nitrogen chain (NO, NO_2) which is believed to be intimately involved in the regulation of the present level of O_3 in the atmosphere[7-10]. At stratospheric temperatures, ClO reacts with O six times faster than NO_2 reacts with O (refs 11, 12). Consequently, the Cl-ClO chain can be considerably more efficient than the NO·NO_2 chain in the catalytic conversion of $O_3 + O \rightarrow 2O_2$ per unit time per reacting chain[12].

Photolytic sink

Both $CFCl_3$ and CF_2Cl_2 absorb radiation in the far ultraviolet[4], and stratospheric photolysis will occur mainly in the 'window' at 1,750–2,200 Å between the more intense absorptions of the Schumann–Runge regions of O_2 and the Hartley bands of O_3.

Figure 8.2. First page of the article by Molina and Rowland published by the journal *Nature*. Reproduced from Molina, M. J., and F. S. Rowland, *Nature* 249 (1974), 810–12.

Molina and Rowland showed that this source exists and results from the emission into the atmosphere of chlorofluorocarbons (CFCs). These nonflammable, noncorrosive, colorless, and odorless organic compounds, and in particular two of them, the CFC-11 ($CFCl_3$) and CFC-12 (CF_2Cl_2), were produced by industry, mainly in the United States, Europe, and Japan. They were used in large quantities since the 1960s in various applications: propellants in aerosol cans, cooling agents in air conditioning systems, solvents, foam blowing agents, etc. In 1974, the world production of CFC-11 was estimated to be 370,000 tons and the CFC-12 production to 440,000 tons. The two Irvine researchers therefore highlighted the existence of anthropogenic products that, after being released into the atmosphere, would directly threaten the ozone layer.

Sherry Rowland—as he called himself—studied at the University of Chicago where he attended courses taught by such prestigious personalities as Enrico Fermi (1901–1954), the inventor of the nuclear reactor and the architect of the atomic bomb, and Harold Urey (1893–1981), a leading isotope specialist who discovered deuterium. Both are Nobel Prize laureates. Urey was interested in the Earth's primitive atmosphere and believed that it was

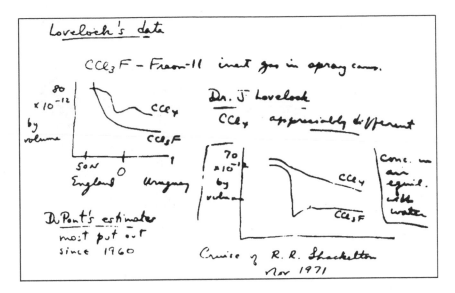

Figure 8.3. Handwritten notes by Sherry Rowland taken during Lester Machta's presentation at the Fort Lauderdale symposium in February 1972.

essentially composed of ammonia, methane, and hydrogen. The appearance of life would have resulted from reactions between these molecules in the presence of electric discharges produced by thunderstorms and leading to the formation of amino acids.

It is therefore not surprising that for much of his career, Rowland was interested in the role of radioisotopes such as carbon-14 in the atmosphere. In February 1972 in Fort-Lauderdale, Florida, he attended a workshop on carbon compounds in the atmosphere. Lester Machta (1919–2001), a meteorologist with the National Oceanic and Atmospheric Administration (NOAA), presented measured concentrations of CFC-11 over the Atlantic Ocean (Figure 8.3). These observations had been made aboard the polar research vessel Shackleton by the renowned scientist James Lovelock (Figure 8.4).[2] Lovelock, the author of the "Gaia hypothesis," which postulates that the Earth functions as a self-regulating system, did not appear to be particularly concerned by the presence of this gas in the atmosphere. In fact, he wrote in an article published by the journal *Nature* on January 19, 1973, that these

---

2. CFC-11 was measured by an electron capture detector invented by Lovelock in 1957. This device detects atoms and molecules in a gas through the attachment of electrons (electron capture ionization).

Figure 8.4. (Left) Frédéric Swarts who synthesized chlorofuorocarbon-12 for the first time at the University of Ghent in Belgium. Source: Timmermans, J., and R. E. Oesper, Frederic Swarts (1866–1940), *J. Chem. Educ.* 38 (1961), 423. (Right) James Lovelock, The initiator of the Gaia hypothesis, who measured for the first time the concentration of chlorofluorocarbons in the atmosphere for the first time. Source: https://www.thextraordinary.org/james-lovelock.

compounds can accumulate in the atmosphere but present no danger.[3] The measurements of Lovelock showed that the atmospheric concentrations of the CFCs are highest in the northern hemisphere, which seemed to confirm that the gas is of industrial origin. Rowland therefore was convinced that CFCs are a threat to the environment and decided to further study this question.

It was at the University of Ghent in Belgium at the end of the nineteenth century that, for the first time, the CF_2Cl_2 molecule (CFC-12)[4] was synthesized. The project was conducted by chemist Frédéric Swarts (1866–1940) (Figure 8.4), a professor in the civil engineering group of the University. Swarts' work prompted Thomas J. Midgley Jr. (1889–1944) and his laboratory assistants at the Thomas and Hochwalt Laboratories[5]

---

3. Lovelock writes: "During the past few decades, the production of the chlorofluorocarbons, the propellant solvents for aerosol dispensers, has grown exponentially. [...] They are unusually stable chemically and only slightly soluble in water and might therefore persist and accumulate in the atmosphere [...] The presence of these compounds constitutes no conceivable hazard."

4. Molecule named dichlorodifluoromethane and commercialized as Freon-12.

5. Which became a research division of Monsanto Chemical Company.

in Dayton, Ohio, Albert L. Henne (1901–1967) and Robert R. McNary (1903–1988), to develop a refrigerant based on CFCs in 1928. Midgley, who was also the inventor of lead-enriched gasoline, was a strong proponent of the use of CFCs in various applications. At a meeting of the American Chemical Society in April 1930, he demonstrated the nontoxic and nonflammable nature of CFCs by inhaling the substance and then exhaling it over a lit candle, and extinguishing its flame. As early as 1931, the CFC-12 (CF_2Cl_2) was industrially produced by *DuPont* under the trade name *Freon*. It was adopted in 1934 by the company *Frigidaire*, which used it in its refrigerators. Two years earlier, CFC-11 was introduced as a fluid in air conditioning systems. Finally, the use of CFC-11 and CFC-12 as propellants in aerosol cans, particularly for cosmetic products, appeared in 1943. Other chlorofluorocarbons were produced by the industry. Organic molecules containing bromine, called *halons*, were later synthesized and used in extinguishers. Bromine is considerably more effective at destroying ozone than chlorine.

Interestingly, from August 16–20, 1971, one year before Lovelock reported the first atmospheric observations of chlorofluorocarbons in the atmosphere, and three years before the seminal publication of Molina and Rowland's paper, NASA convened a working group at its Center in Langley, Virginia, to examine how to monitor air and water pollution from space. A Panel on "Gaseous Pollutants and Natural Trace Gases" chaired by William Kellogg from NCAR (National Center for Atmospheric Research) highlighted for the first time that chlorofluorocarbons released in the atmosphere could be detrimental for the environment (Figure 8.5):

Fluorocarbons such as the Freon are considered as relatively harmless. [...] The inertness of these compounds, which is their virtue in most applications, is also the cause for environmental concern. In addition to being chemically inert, the fluorinated compounds are optically inert. Photo-dissociation requires very short ultraviolet wavelengths, which are available only at the highest levels of the atmosphere. At present there does not seem to be any known mechanism for removal of this class of fluorinated compounds from the atmosphere. Accumulation can therefore be expected to continue, resulting in an apparently irreversible alteration of the atmospheric composition. Clearly, this situation is cause for concern. It is recommended that the physical and chemical properties of the fluorinated compounds be studied further and their atmospheric concentrations be monitored.

At present the concentrations of these compounds are not likely to be greater than the parts-per-trillion level. Such concentrations are detectable by the sensitive electron-capture technique, but are well below the detectability level of space-borne optical systems. However, the fluorocarbons have distinctive strong infrared bands, and can be expected to become detectable optically when their concentrations approach the parts-per-billion level. A measurement accuracy of better than 0.001 ppb may be required.

After having heard the presentation of Lester Machta in Fort-Lauderdale, Rowland became convinced that chlorofluorocarbons represented an important anthropogenic source of chlorine in the stratosphere. He hired a Mexican "postdoc," Mario Molina, who had just completed his thesis at Berkeley, to work on this question. Mario José Molina-Pasquel Henriquez, whose father was one of Mexico's high-profile ambassadors, acquired a basic education first in Mexico City and then in a private school in Switzerland before completing his university studies in Mexico, Germany, and in the

Fluorocarbons

Volatile fluorocarbons, of which freon is perhaps the best known, are released in vast quantities into the atmosphere and must remain there as virtually permanent components of the atmosphere—at least we know of no removal mechanisms. The question naturally arises as to whether this continual release will create a problem as they continue to build up. No remote methods for the measurement of these compounds could be identified.

Figure 8.5. Report on remote measurement of pollution produced in 1971 by the Langley Research Center of NASA and highlighting for the first time the unknown but possibly large impact of fluorocarbons (also called chlorofluorocarbons) on the environment. Source: National Aeronautics and Space Administration (NASA-SP 285).

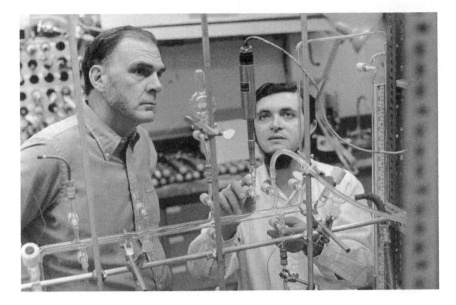

Figure 8.6. Sherry Rowland (left) and Mario Molina (right) in their laboratory at the University of California at Irvine. Both are recipient of the 1995 Nobel Prize for Chemistry for demonstrating that chlorofluorocarbons could destroy part of the ozone layer. Source: https://undsci.berkeley.edu/article/cfcs_checklist.

United States. He joined Sherry Rowland's group at the University of California at Irvine in 1973 and decided to study the processes leading to the slow degradation of chlorofluorocarbons in the atmosphere. These compounds have been in fact injected in the atmosphere in increasingly large quantities, and because they are chemically very stable, their life expectancy in the atmosphere reaches several decades. In their article published on June 28, 1974, in *Nature*, Molina and Rowland (Figure 8.6) stated that the concentration of CFCs was expected to increase significantly in future years. Dissociation of these molecules by ultraviolet solar radiation in the upper stratosphere was expected to release significant amounts of chlorine that would destroy much of the ozone layer.

This alarming study was followed by a detailed investigation conducted by the scientific community, the conclusions of which were published by the US Academy of Sciences. Subsequently, in 1978, the U. S. government decided to ban the use of CFCs as propellant in aerosol cans. Other countries adopted even more stringent regulations, and, for example, in 1980, the Parliamentary Assembly of the Council of Europe in Strasburg, France, voted a resolution inviting governments to promote scientific research aimed at evaluating the effects on ozone of chlorofluorocarbons and at developing substitute products.[6] On March 22, 1985, an International Convention for the Protection of the Ozone Layer was signed in Vienna. It was ratified by 197 states. In addition to this Convention, a Protocol on substances that destroy ozone was also being established. This international treaty was the subject of intense political discussions between nations under the aegis of the United Nations. The lead role was played by the United States, and particularly by the head of the US delegation, Ambassador Richard E. Benedick (Box 8.1), who showed great skill and determination. Several countries, notably the Scandinavian countries, were in favor of measures to protect the ozone layer, while others, particularly CFC-producing countries such as the United Kingdom, France, and Italy were more reluctant. The industry was originally opposed to regulating the CFCs, but in a paid advertisement published by the *New York Times* in 1975, the *DuPont* company stated,

> Should reputable evidence show that some fluorocarbons cause a health hazard through depletion of the ozone layer, we are prepared to stop production of the offending compounds.

Finally, the protocol was signed in Montreal on September 16, 1987 by 46 signatories, and was later ratified by all UN members as well as Nuie, the Cook Islands, the Holy See,[7] and the European Union. It was supplemented at a later stage by a series of amendments that further limited and ultimately banned the production of the main substances responsible for the ozone destruction. This result was a victory for Molina and Rowland who shared with Paul Crutzen the 1995 Nobel Prize for Chemistry. It was regarded as the successful development of a new form of diplomacy, sometimes named "environmental diplomacy." It was probably facilitated by the fact that substitutes for the chlorofluorocarbons had become available.

Rowland who raised the issue of chlorofluorocarbons and their effect on the ozone layer continued his research at the University of California at Irvine. Despite the attacks from industry that he had to face, he felt during his career that he had to inform decision-makers of the consequences of

---

6. See report (Doc 4558 of the Committee on Science and Technology, Guy Brasseur Rapporteur) adopted on July 3, 1980.

7. The Holy See is a universal ecclesiastical jurisdiction of the Catholic Church and a sovereign entity in international law.

Box 8.1. Ambassador Richard Elliot Benedick and global diplomacy

Richard Benedick, the chief US negotiator of the Montreal Protocol on the protection of the ozone layer is regarded as one of the principal architects of this landmark international agreement. As a career diplomat, Benedick served in Iran, Pakistan, France, Germany, and Greece, and as Deputy Assistant Secretary of State for Environment, Health, and Natural Resources, he supervised international negotiations on climate change, stratospheric ozone, biotechnology, tropical forests, oceans, and wildlife conservation. His book *Ozone Diplomacy: New Directions in Safeguarding the Planet* (Harvard University Press, 1998) highlights the efforts that led to the signature of the Montreal Protocol hailed at the time as "the most significant international environmental agreement in history" and a successful example of what is now called global diplomacy. In his book, Benedick states that "the problem of protecting the stratospheric ozone layer presented an unusual challenge to diplomacy. Military strength was irrelevant to the situation. Economic power also was not decisive. Neither great wealth nor sophisticated technology was necessary to produce large quantities of ozone-destroying chemicals. Traditional notions of national sovereignty became questionable when local decisions and activities could affect the well-being of the planet. Because of the nature of ozone depletion, no single country or group of countries, however powerful, could effectively solve the problem. Without far-ranging cooperation, the efforts of some nations to protect the ozone layer would be undermined."

Highlighting the role of scientists in the process, Benedick stated in his testimony to the Committee of Environment and Public Works of the United States Senate on September 28, 2005, "Science became the driving force behind ozone policy, but it was not sufficient for scientists merely to publish their findings. In order for the theories to be taken seriously and lead to concrete policies, scientists had to interact closely with government policy makers and diplomatic negotiators. This meant that they had to leave the familiar atmosphere of their laboratories and assume an unaccustomed shared responsibility for the policy implications of their research. The history of the Montreal Protocol is filled with instances of scientific panels being called upon to analyze and make informed judgments about the effectiveness and consequences of alternative remedial strategies and policy measures. [...] Politics, stated Lord Kennet during early ozone debates in the House of Lords, is the art of taking good decisions on insufficient evidence. The success of the Montreal Protocol stands as a beacon of how science can help decision makers to overcome conflicting political and economic interests and reach solutions.

The ozone history demonstrates that even in the real world of ambiguity and imperfect knowledge, the international community, with the assistance of science, is capable of undertaking difficult and far-reaching actions for the common good."

Figure 8.7. Ambassador Richard E. Benedick. Source: http://enb.iisd.org/ozone/mop19/anniversary.htm.

human activities for the environment. At a meeting on climate change held in 1997 at the White House, Rowland asked the question[8]:

Is it enough for a scientist simply to publish a paper? Isn't it a responsibility of scientists, if you believe that you found something that can affect the environment, to actually do something about it, enough that action actually takes place? If not us, who? If not now, when?

Molina joined the Massachusetts Institute of Technology near Boston and later the Scripps Institution of Oceanography at La Jolla on the California coast before returning to Mexico City to direct the Mario Molina Centre for Energy.

International agreements on ozone and specifically the "Montreal Protocol on Substances that Deplete the Ozone Layer" have greatly benefited from the input of the science, which highlighted the vulnerability of the ozone layer, and produced scenarios showing future ozone depletion in

8. Quoted by Donald R. Blake and Isobel J. Simpson in EOS, 93, October 9, 2012.

the stratosphere if immediate safeguarding measures were not adopted. To facilitate the transfer of information to decision makers and in particular to negotiators of international agreements, the results of research published by the scientific community in peer-reviewed journals were critically analyzed, evaluated, and summarized in reports published periodically by the United Nations Environment Programme (UNEP) in Nairobi together with the World Meteorological Organization (WMO) in Geneva (Figure 8.8). These assessment reports were written by a group of several dozen specialists and were reviewed and examined by many independent experts. The reviews were addressed by the authors who traditionally met for the occasion in the quaint village of Les Diablerets in Switzerland. Leading coordinators of the earlier reports were Daniel Albritton, director of the NOAA Aeronomy Laboratory in Boulder, Colorado, Gérard Mégie (1946–2004), director of the "Service d'Aéronomie du CNRS" in France before becoming president of the CNRS in Paris and Sir Robert Watson, then a senior scientist at NASA in Washington, District of Columbia (Figure 8.9). Albritton was leading a team of outstanding scientists who played a key role in the investigation of the stratosphere and of the processes

Figure 8.8. Several early periodic scientific assessments of stratospheric ozone including the 1988 report by the International Ozone Trend Panel. Source: World Meteorological Organization, Geneva, Switzerland.

Figure 8.9. (From left to right) The coordinators of various international ozone assessments: Daniel Albritton (courtesy: National Oceanic and Atmospheric Administration), Robert Watson (reproduced from https://www.nature.com/articles/d41586-018-05984-3), and Gérard Mégie (courtesy: Centre National de la Recherche Scientifique, France).

that led to ozone depletion. He was exceptionally skilled in communicating complex scientific information to policy-makers and convinced NOAA to be engaged at the forefront of ozone research. Mégie, a former student of Ecole Polytechnique in Paris and a member of the French Academy of Sciences, developed original methods for measuring ozone and other atmospheric quantities by laser sounding. Watson, born and educated in the United Kingdom and a former postdoctoral Fellow in Harold Johnston's laboratory, was a specialist of chemical kinetics who moved to the Jet Propulsion Laboratory in Pasadena, California, in 1976. He joined NASA Headquarters in 1979 where he shaped the international stratospheric research for several years, and initiated, for example, a detailed assessment of post-1970 ozone trends in the stratosphere (see Chapter 9). His major accomplishment has been to set up the agenda that led to the gradual phase-out of chlorofluorocarbons. Watson was during several years the Director of NASA's Stratospheric Ozone Program and, during the Clinton Administration, worked closely with Vice President Al Gore as the Associate Director of Environment at the White House Office of Science and Technology Policy.

Ozone assessments (Fig. 8.10) were considered as representing the scientific consensus by the international research community, and their credibility was such that they served as an undisputed scientific reference on the ozone issue and were essential input to international agreements. They also constituted a model for the upcoming reports on climate change published later by the Intergovernmental Panel for Climate Change (IPCC).

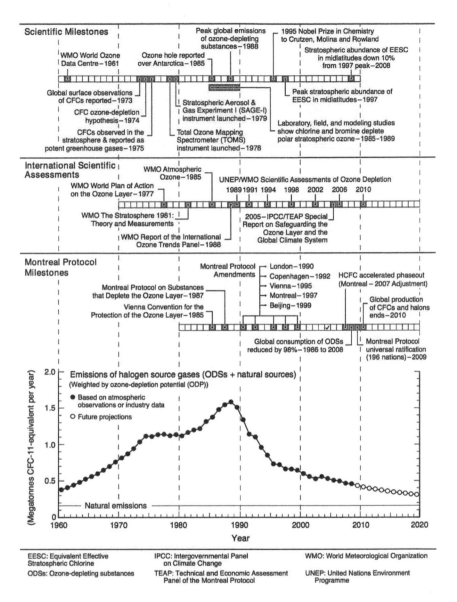

Figure 8.10. Scientific milestones, scientific assessments, and international agreements related to the Montreal Protocol on Substances that Deplete Ozone represented as a function of time. The lowest panel shows the evolution of the atmospheric load of anthropogenic chlorine as a function of time expressed as the mass of CFC-11 equivalent (Megatons per year). The data refer to observations for the period 1960 to 2009 and projections in the following years. Reproduced from Fahey, D. *Theoretical Inquiries in Law 14* (2013), 21–42.

CHAPTER NINE

The Antarctic Ozone Hole

The Discovery of the Ozone Hole

Since the International Geophysical Year of 1957 to 1958, ozone has been measured systematically above the Antarctic continent. This has been particularly the case at the British Polar Station at Halley Bay, where a Dobson spectrophotometer was installed and has been performing routine measurements for many years. Since 1976, physicist Joseph Charles (Joe) Farman (1930–2013), Head of Physics Section at the British Antarctic Survey at Cambridge, had been responsible for these ozone measurements with his colleagues Brian Gardiner and Jonathan Shanklin (Figure 9.1). The ozone column that the team observed appeared to decrease from year to year during the austral spring (September, October). Farman feared that the old device used at Halley bay was recording a drift due to the malfunctioning of the instrument. What seemed curious to him was that National Aeronautics and Space Administration (NASA), whose satellites had been constantly monitoring the atmosphere and measuring ozone, had not reported such a decrease. Only a meteorologist working for the Japanese Meteorological Agency in Tsukuba, Japan, Shigeru Chubachi, had indicated during a scientific conference held in 1984 in Greece that ozone levels had been very low in September and October 1982 at the Japanese Antarctic station of Syowa (Figure 9.2).

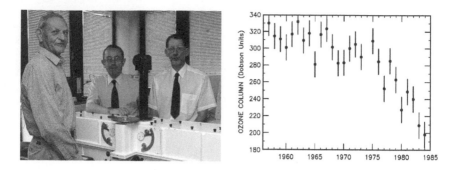

Figure 9.1. (Left) Joe Farman, Brian Gardiner, and Jonathan Shanklin at the British Antarctic Survey in front of an ozone spectrophotometer. (Right) Evolution of the ozone column (Dobson units) measured at Halley Bay (76°S) in Antarctica during the month of October 1957 to 1984. Adapted from Farman et al., *Nature* 315 (1985), 207–10.

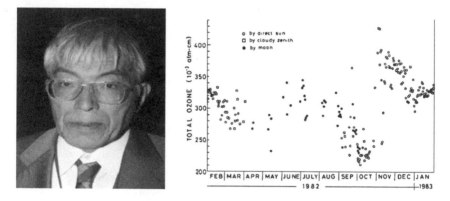

Figure 9.2. The exceptionally low values of the total ozone column observed at the Antarctic station of Syowa during the months of September and October 1982 (right panel). The data were presented by the Japanese scientist Shigeru Chubachi (left panel) at the Quadrennial Ozone Symposium organized at Halkidiki in northern Greece by the International Ozone Commission during the summer 1984. Ozone balloon soundings made at the same station show that the ozone concentrations are particularly low in the lower stratosphere. Reproduced from Chubachi, S. "A special ozone observation at Syowa station, Antarctica from February 1982 to January 1983." In *Atmospheric ozone*, eds. C. S. Zerefos and A. Ghazi (Norwell, MA: D. Reidel, 1985), 285–89.

Farman, a native of the Norwich region in England, studied at Corpus Christi College in Cambridge before joining the aeronautics firm Haviland to begin his professional career. However, the young researcher was attracted to the adventure of polar expeditions and in 1956 joined the Falkland Island

Dependency Survey (which later became the British Antarctic Survey) to devote many years of his scientific career to environmental problems in this remote region of the world.

Before publishing his observations on ozone in the early 1980s, Farman took a few precautions. First, he installed at Halley Bay a second spectropho-tometer and verified that it measured ozone values almost identical to those determined by the older instrument. Then, on October 10, 1983, Shanklin contacted the manager of ozone measurements at NASA's Wallops Center (Figure 9.3). He asked him by letter whether his agency could confirm the abnormally low concentrations measured at Halley Bay and whether the NASA scientists attributed the ozone depletion to the effects of the 1982 eruption of the Mexican volcano El Chichon. The response from the Head of the Balloon Project Branch at the Wallops Centre was short, bureaucratic, and thus of little use: "Our group is no longer involved in this activity." Shanklin had contacted the wrong group at NASA.

Farman then decided to move his ozone instrument to another part of Antarctica, 1,500 km northwest of Halley Bay, where he also observed very small amounts of ozone during the southern spring. Despite the neg-ative opinion of the head of his department, who feared that the British Antarctic Survey would be embarrassed if the study's conclusions turned out to be incorrect, Farman decided to publish his results in *Nature* (Figure 9.4). The article appeared in May 1985 and stated that the ozone depletion observed in October (almost 40% in ten years) could be attrib-uted to chlorine compounds of industrial origin. No convincing chemical mechanism, however, was presented to explain this drastic springtime ozone reduction. Farman, a kind person who enjoyed smoking the pipe while chatting with his colleagues, realized the importance of his discovery, and became a strong advocate for the need to protect the Earth's environment. He believed in the importance of systematic measurements and liked to stress that "too much money is going into expensive climate modelling computers, and not enough into basic observational science."

The initial reactions to Farman's article by his colleagues from the scientific community were colored by skepticism, but raised a number of challenging scientific questions. The mathematical models showed that reactive chlorine should destroy only a tiny fraction of stratospheric ozone, and that this effect should have manifested itself at all latitudes around 45 or 50 km altitude. In fact, in 1983, only two years before Farman's paper appeared in *Nature*, a report by the US National Research Council projected that the use of chorofluorocarbons (CFCs) at then-current rates

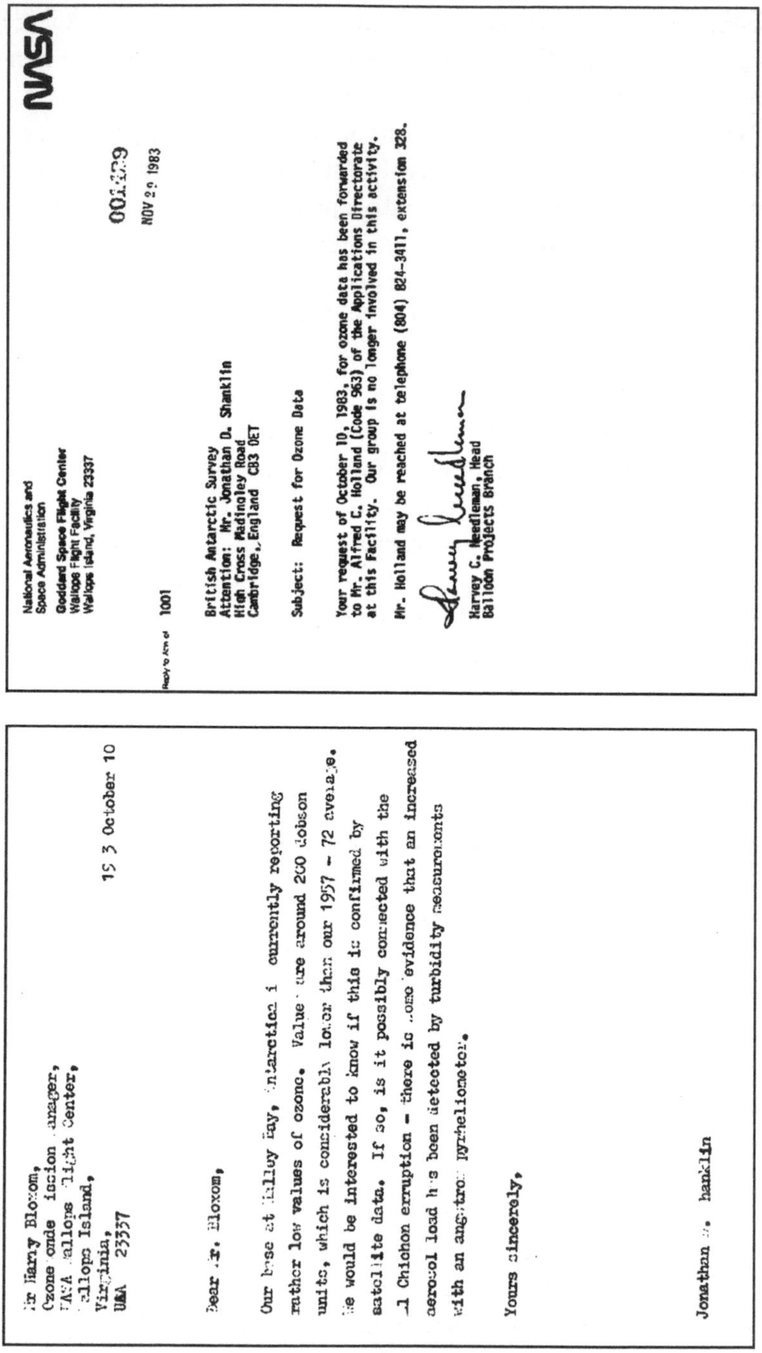
Mr Harry Bloxom,
Ozonesonde Mission Manager,
NASA Wallops Flight Center,
Wallops Island,
Virginia,
USA 23337

1983 October 10

Dear Mr. Bloxom,

Our base at Halley Bay, Antarctica is currently reporting rather low values of ozone. Values are around 200 dobson units, which is considerably lower than our 1957 – 72 average. We would be interested to know if this is confirmed by satellite data. If so, is it possibly connected with the El Chichon erruption – there is some evidence that an increased aerosol load has been detected by turbidity measurements with an angstrom pyrheliometer.

Yours sincerely,

Jonathan D. Shanklin

National Aeronautics and Space Administration
Goddard Space Flight Center
Wallops Flight Facility
Wallops Island, Virginia 23337

NASA

001429

NOV 29 1983

Reply to Attn of: 1001

British Antarctic Survey
Attention: Mr. Jonathan D. Shanklin
High Cross Madingley Road
Cambridge, England CB3 0ET

Subject: Request for Ozone Data

Your request of October 10, 1983, for ozone data has been forwarded to Mr. Alfred C. Holland (Code 963) of the Applications Directorate at this Facility. Our group is no longer involved in this activity.

Mr. Holland may be reached at telephone (804) 824-3411, extension 328.

Harvey C. Needleman, Head
Balloon Projects Branch

Figure 9.3. Exchange of letters between Jonathan Shanklin from the British Antarctic Survey and NASA staff at the Wallops Flight Center. Reproduced from RealClimate.org.

Large losses of total ozone in Antarctica reveal seasonal ClO$_x$/NO$_x$ interaction

J. C. Farman, B. G. Gardiner & J. D. Shanklin

British Antarctic Survey, Natural Environment Research Council, High Cross, Madingley Road, Cambridge CB3 0ET, UK

Recent attempts[1,2] to consolidate assessments of the effect of human activities on stratospheric ozone (O_3) using one-dimensional models for 30° N have suggested that perturbations of total O_3 will remain small for at least the next decade. Results from such models are often accepted by default as global estimates[3]. The inadequacy of this approach is here made evident by observations that the spring values of total O_3 in Antarctica have now fallen considerably. The circulation in the lower stratosphere is apparently unchanged, and possible chemical causes must be considered. We suggest that the very low temperatures which prevail from midwinter until several weeks after the spring equinox make the Antarctic stratosphere uniquely sensitive to growth of inorganic chlorine, ClX, primarily by the effect of this growth on the NO_2/NO ratio. This, with the height distribution of UV irradiation peculiar to the polar stratosphere, could account for the O_3 losses observed.

Figure 9.4. The discovery of the Antarctic ozone hole: Summary of the article published by Farman et al. in *Nature*. The paper highlighted the presence of a large ozone destruction during the months of September and October. Reproduced from Farman et al., *Nature* 315 (1985), 207–210.

would lead to depletion of the total global ozone layer by only about 3% in about a century. In other words, the models could not account for the large decrease in the ozone column observed at Halley Bay. Further, they could not explain the high and rapid ozone erosion occurring only in the Antarctic region. Another crucial piece of information soon came from balloon surveys conducted from the Amundsen-Scott base at the South Pole. The measurements showed that ozone had virtually disappeared in an altitude range of 16 to 26 km, and that no significant ozone depletion appeared to occur outside these specific heights. It was clear that something dramatic was suddenly occurring in the stratosphere above the Antarctic continent, something that nobody had expected and that nobody could explain.

Space Observations of Ozone Depletion

An important piece of information was missing. NASA that was constantly monitoring ozone from space had not reported any substantial ozone depletion, and therefore one could have doubts about Farman's measurements. In fact, already in October 1983, thus before the British team had released its dramatic observations, the NASA Ozone Processing Team in charge of the ozone observations by the TOMS (Total Ozone Mapping Spectrometer) and SBUV (Solar Backscatter Ultraviolet Radiometer) space instruments had noticed abnormally low ozone concentrations in Antarctica, and was exploring the causes of these anomalous values. Was it an instrument's artifact or was it a real ozone erosion? One approach to address this question was to compare the space data with surface measurements, generally referred to as "ground truth." The NASA team rapidly noticed that the TOMS values were about a factor of two lower than the measurements made by the Dobson spectrophotometer installed at the Amundsen-Scott station at the South Pole. It concluded therefore that their surprising measurements were extremely suspicious. While trying to understand the discrepancy, they wrote in an internal memorandum on October 1, 1984:

> Anomalously low ozone over Antarctica in October of 1983: Instrumental problems and clouds have been ruled out as causes. Though TOMS also measures low ozone, the Amundsen-Scott station does not during this period. A possible explanation could be a short-lived anomaly in the spacecraft attitude.

As long as the discrepancy was not explained, the satellite data were not publicly released. Several actions, however, were considered: Rule out the possibility of an error in the spacecraft attitude; study continuous scan for anomalies; get data from the ground station at Syowa; process and archive the suspicious observations, but warn users. The discrepancy, however, was finally resolved when it was established that due to a mistake made by the operator of the instrument, the spectrophotometer data at the South Pole station were erroneous and uncorrectable.[1] Thus, the satellite observations were eventually judged to be representative of reality and were released (Box 9.1).

---

1. Observations using the Dobson instrument were made by using incorrect pairs of wavelengths to derive the ozone column abundance.

Box 9.1. Did NASA overlook the ozone hole?

The announcement by British scientists of a massive and unexpected ozone destruction over the Halley Bay station in Antarctica during the austral spring, and specifically the abnormally low values of the ozone column recorded in October 1983 landed like a bomb shell. The question was immediately asked: why didn't NASA that was operating the Nimbus 7 satellite (TOMS, SBUV) report the existence of this dramatic geophysical event? Had the agency been negligent and did it fail in its mission?

A response came rapidly: the quantities detected by the satellites were so small compared to the climatological values that NASA considered them as suspicious and hence flagged them or even rejected them. Sherry Rowland supported this point of view in his book "The CFC Ozone Puzzle" by stating about NASA's scientists, "What they did was program their data to reject, but notify, that some unusually low ozone was being recorded" and Robert Watson, then Head of the NASA's Upper Atmosphere Program, who was interviewed in the film entitled "Ozone hole: How we saved the Planet" added: "We went back and looked carefully at the satellite observations and we realized that the way we were processing the data was wrong. If you get strange results, you just reject them." This point of view implies that it is the publication by Farman in May 1985 that led the embarrassed agency to reexamine its data.

And yet, is this the right explanation? Is the NASA's alleged negligence simply a myth circulating among scientists and journalists? Or was NASA aware of the existence of the ozone hole before the British researchers published their findings?

In fact, the analysis of the October 1983 observations started in August 1984. The team responsible for this task at the NASA Goddard Space Flight Center was led by Albert J. Fleig and included P. K. Bhartia and Donald Health. The existence of low ozone-column values over Antarctica was soon recognized, but the data were considered to be suspect, and perhaps degraded by a malfunction of some sensor onboard the satellite. NASA therefore decided to compare its space observations with measurements made by the Dobson spectrophotometer deployed at the US Amundsen-Scott research station located at the South Pole. Fleig and his colleagues found with great surprise that the ozone columns measured by this ground-based instrument during October 1983 were of the order of 365 Dobson units, thus well above the values of about 185 Dobson units deduced from the space observations. They concluded that the data provided by the satellites were probably incorrect, and certainly suspicious. They decided therefore to delay the release of these observations to the public. What the NASA team failed to do, however, was to compare their data with the observations made in October 1983 at the Japanese Syowa polar station; these observations had been presented in August 1984 at the Quadrennial Ozone Conference in Greece.

The internal evaluation of NASA data continued throughout 1984, and in December the Fleig group became confident that the measurements from Nimbus 7 were in fact correct. The team therefore decided to submit an abstract of a presentation to the upcoming IAGA scientific assembly to be held in Prague from August 5–17, 1985. The summary sent to the organizers of the meeting by Bhartia in December 1984 (five months before the publication of Farman's study) indicated that "a sharp decrease in ozone density in the south polar stratosphere was observed by the SBUV/TOMS instruments on the Nimbus-7 satellite in October of 1983 and 1984. The measured total column ozone levels dropped as low as 0.150 atm-cm (a value not recorded elsewhere) but recovered to their normal values within a month." 0.150 atm-cm is equivalent to 150 Dobson units.

During 1985, NASA started to release its polar ozone data, and in November 1985, Walter Sullivan published an article in the *New York Times* (Figure 9.6) that showed for the first time a satellite view of what the journalist called "a hole" in the ozone layer, a name coined a bit earlier by Sherry Rowland. In April 1986, almost a year after Farman's publication, Richard Stolarski (NASA/ Goddard Space Flight Center) submitted to *Nature* a detailed analysis of ozone observations by TOMS and SBUV. The paper was published in August 1986. Stolarski likes to repeat that the original title of his article included the words "ozone hole," but the reviewers asked him not to use this term considered to be somewhat unscientific.

At that time, it remained to understand why the spectrophotometer installed at the South Pole station had not observed the low ozone values during the 1983 spring time. In November 1986, Walter D. Komhyr, a NOAA ozone specialist, provided the answer: "Data previously reported for October-December 1983 have been identified as erroneous and uncorrectable (observations were incorrectly made on the so-called A', C' and D' rather than on A, C and D wavelengths)."

In conclusion, NASA did not overlook the ozone hole; however, it took the agency a long time to retrieve, analyze, evaluate, and eventually release the observations of the massive Antarctic ozone erosion. In 2018, more than 30 years after the discovery of the ozone hole, Bhartia, who was directly involved in the retrieval of the Nimbus 7 measurements referred to the difficulties encountered by the agency. He used the expression "Black Swan" coined by the Lebanese-American essayist writer Nassim Taleb (also a statistician and a specialist in risk assessments) to describe a highly improbable event that comes as a surprise and has a major impact on the discipline in which it occurs. Such events are observed, said Bhartia, in the fields of science, religion, culture, and politics. The discovery of the ozone hole was certainly a black swan for the environmental sciences. Referring more specifically to the role of NASA in the release of information on the ozone hole, Bhartia highlighted "the chaos that happens when black swan events occur and how misinformation and theories are created post-hoc to explain what did or did not occur." And he concluded, "The key lessons to be learned from this experience is the continuing need to collect high-quality data from in situ instruments to generate a priori profiles for satellite algorithms and validation data to confirm satellite findings. An appropriate metaphor is that satellites expand the reach of in situ measurements as a surgeon using robotic instruments does for patients in remote areas. However, as robotic instruments cannot replace the surgeon, satellite instruments often do not replace the capabilities of in situ instruments."

The space-based measurements left no doubt that there was a *hole* in the ozone layer. The first satellite image of the ozone hole (Figure 9.5) was publicly presented by Pawan Kumar Bhartia (Figure 9.7) of NASA's Goddard Space Flight Center at a symposium organized in Prague, Czechoslovakia, by the International Aeronomy and Geophysics Association (IAGA) in August 1985. Subsequent observations released by the US Space Agency showed that

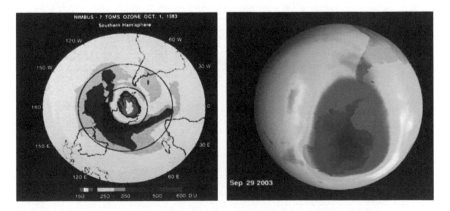

Figure 9.5. (Left panel): The first satellite image of the ozone hole observed by the TOMS instrument on October 1, 1983, and presented by scientist Pawan K. Bhartia from NASA Goddard Space Flight Center at the 5th *IAGA* Scientific Assembly in *Prague* (August 1985). Reproduced from RealClimate.org. (Right panel): The ozone hole 20 years later, on September 29, 2003. From NASA, https://svs.gsfc.nasa.gov/2989.

Low Ozone Level Found Above Antarctica

By WALTER SULLIVAN

Satellite observations have confirmed a progressive deterioration in the earth's protective ozone layer above Antarctica, according to scientists who analyzed data recently sent back from space.

Each October, the data show a "hole" appears in the ozone layer there, scientists say, and each year the layer in that area becomes less able to shield the earth from damaging solar ultraviolet rays.

Since 1974 scientists have been predicting a gradual depletion of stratospheric ozone as a result of increased pollution of the atmosphere. The new data have persuaded some researchers that the ozone loss is proceeding much faster than expected.

Link to Skin Cancer

It has been predicted that a significant depletion of the ozone layer would substantially increase the rate of skin cancer worldwide. Even under normal conditions, however, the ozone layer is subject to wide variations, and whether the recent depletion is part of a long-term trend is difficult to establish.

Several substances introduced into the atmosphere as pollutants are suspected of contributing to the depletion, chief among them fluorocarbons, such as the Freon used for refrigeration, and methane, nitrous oxide and a variety of bromine compounds.

The satellite measurements indicating a rapid decrease over Antarctica have been made by two devices riding the Nimbus-7 satellite, which was launched in 1978. Dr. Donald F. Heath of the Goddard Space Flight Center in 'Greenbelt, Md., who for several years has been monitoring the recordings, said yesterday a quick look at last month's data indicated that the decline is continuing.

In his view, however, the reason for it remains uncertain. It was first blamed on sulfur compounds and other particles ejected into the stratosphere by the 1982 eruption of El Chichon in Mexico.

Scientists Backs Theory

This explanation was also advanced by H. U. Dütsch of the Federal Institute of Technology in Zurich, Switzerland, based on ozone measurements at Arosa in the Swiss Alps.

The measurements there, as at numerous other ground stations, are based on recording two wavelengths of sunlight. Ozone absorbs sunlight at one of the wavelengths, so the relative strength of the two wavelengths is an indication of how much of the gas is in the atmosphere. The 1983 average was the lowest in 60 years. If that was entirely caused by material from El Chichon, Dr. Rowland said in a recent interview, the level should now be returning to normal, but it is not.

According to Dr. Heath, however, there are other possible explanations. The decrease could be linked to the sunspot cycle, which is now near a minimum. According to a study by NASA scientists, the chemical reactions that produce stratospheric ozone are stimulated by a form of ultraviolet radiation that becomes weak when sunspots are fewest.

Unusual Conditions Noted

Nor is it clear, Dr. Heath said, whether the Antarctic readings manifest a local change in atmospheric circulation, rather than a global depletion. The condition of the winter atmosphere over Antarctica is not matched anywhere else. The atmosphere, immersed in the polar night, remains highly stable and becomes extremely cold. Then, when spring comes to the Southern Hemisphere about October, it is suddenly bathed in sunlight and, it is hypothesized, ozone depletion runs at full speed.

According to the report observations at Halley Bay in Antarctica, "Comparable effects should not be expected in the Northern Hemisphere where the winter polar stratospheric vortex is less cold and less stable." The report, published earlier this year in Nature,

Decrease in Ozone over Antarctica

Measurements from the Nimbus 7 satellite have shown a "hole" in the ozone layer over Antarctica. These, recorded on Oct. 4, 1983, and now confirmed, indicate ozone abundances in terms of how deep a layer would be formed by the gas, in centimeters, at normal atmospheric temperature and pressure. In addition to the depleted area near the South Pole, there is a persistent high concentration south of Australia.

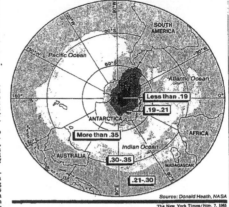

Source: Donald Heath, NASA

The New York Times/Nov. 7, 1985

was by J. C. Farman, B. G. Gardiner and J. D. Shanklin.

That fluorocarbons are responsible for the newly observed depletion of the ozone layer has been proposed by scientists of the British Antarctic Survey, based on observations conducted since 1957 at Halley Bay, and by Dr. F. Sherwood Rowland of the University of California at Irvine. It was Dr. Rowland, Dr. Mario J. Molina and Dr. Harold Johnston who in 1974 first warned of such a danger.

In 1980 a committee of the National Academy of Sciences concluded that the projected ozone depletion, through increased ultraviolet radiation, could increase skin cancer, curtail crop production and destroy the larvae of some marine organisms. A 16 percent ozone reduction, it said, would probably produce each year "thousands" of additional cases of melanoma — the most lethal skin cancer.

Effect of Ban

In 1977 a ban was imposed on fluorocarbons as spray-can propellants, but it became evident that the ozone varies in response to a variety of interacting natural and human influences. By 1984

an academy report estimated ozone reduction, due to fluorocarbons, at only 2 percent to 4 percent.

An annual 20 percent increase in the atmospheric content of bromine compounds that also endanger the ozone layer has been reported by a group from the Max Planck Institute for Aeronomy in Lindau, West Germany. Their instruments were lifted 15 miles above southern France by balloon in the fall of 1982, 1983 and 1984. Production of such compounds, including those used in fire extinguishers, appears to be increasing rapidly.

The original warning by Dr. Rowland and Dr. Molina concerned the chlorine that would be released when fluorocarbons are exposed to ultraviolet rays in the stratosphere. While those synthetic compounds are normally very stable, when exposed to ultraviolet light they break down and one of their constituents is chlorine, which can remove ozone from the atmosphere. The molecules of ozone gas are formed of three oxygen atoms, whereas oxygen gas contains only two of them. When chlorine reacts with an ozone molecule, breaking it up, the chlorine remains intact, ready to attack another one.

Figure 9.6 Article published by the *New York Times* in November 1985.

the hole was gradually expanding from year to year and in the late 1980s covered an area larger than the entire Antarctic continent. The ozone hole extended throughout an entire area (Figure 9.5) surrounded by the vast polar vortex that preserves the region from the dynamical influences of lower latitude regions.

The use of the *ozone hole*, as a powerful metaphor adopted in the context of the convincing visualization material that NASA had distributed to the public, had an overriding impact on the public despite the fact that the scientific community had not yet identified the cause of this phenomenon. It provided a visual evidence that nobody could ignore and probably paved the way to the signing of the Montreal Protocol in 1987. And as Sebastian Vincent Grevsmühl from the Centre de Recherches Historiques at Ecole des Hautes Etudes en Sciences Sociales in Paris states[2]:

> The successful environmental reframing of the "hole" in the sky as a broken shield, letting hazardous ultraviolet rays pass through Earth's broken protective layer, proved therefore a highly influential and efficient image for both, the legislators and the environmental movement. NASA's satellite images which were widely shared in the media as well as the agencies' "ozone hole" animation movies, that showed the temporal evolution of the hole that could become as large as the entire Antarctic continent, rapidly became icons of the precautionary principle as they were shown on national television and at congressional hearings. Indeed, the powerful metaphor and the associated imagery gave support to those groups who believed that the consequences of taking no actions would be far worse than the consequences of over-restrictive regulations.

Despite the rapid progress made by NASA in the analysis of its ozone measurements, there was an urgent need to unambiguously establish if space observations would provide accurate information on possible ozone trends in the stratosphere. The Nimbus 7 satellite with its several instruments measuring ozone, launched in October 1978, had been designed to investigate the dynamics of the ozone layer during a period of only one year, but not to quantify long-term ozone trends. Luckily, the spacecraft remained operational for 14 years, so that some useful information on ozone trends could be

---

2. Grevsmühl, S. V., Revisiting the "Ozone Hole" Metaphor: From Observational Window to Global Environmental Threat, Published in *Environmental Communication*, doi:10.1080/17524032.2017.1371052, 2017.

Figure 9.7. Meeting of ozone measurement specialists convened by Donald Heath in Salzburg, Austria, during the month of August 1985. This meeting whose purpose was to discuss how satellites could detect changes in the abundance of stratospheric ozone, took place only 3 months after the publication of Farman's paper on the ozone hole. Source: https://climate.nasa.gov/news/781/discovering-the-ozone-hole-qa-with-pawan-bhartia/.

derived in the mid-80s. To address this question, Donald Heath, a scientist at NASA, who had initiated the ozone measurement program at Goddard Space Flight Center in the late 1960s, invited several prominent experimentalists to attend an international meeting in Salzburg, Austria (Figure 9.7), in late August 1985, a few days after the IAGA meeting in Prague. The scientists who were present were tasked to examine how to make the best use of space observations to detect possible ozone trends.

Based on the analysis of the TOMS space measurements and its companion SBUV, both on board Nimbus 7, Donald Heath detected a significant ozone depletion at mid-latitudes in the Northern hemisphere. This ozone erosion, however, was substantially larger than the trends deduced from ground-based Dobson instruments. Neil Harris, a British graduate student of Sherry Rowland at Irvine, California, had analyzed ozone measurements made by 22 Dobson instruments located north of 30°N, and had identified a small post-1970 downward ozone trend only during winter time. In order to reconcile different contradictory pieces of information, NASA, under the leadership of Robert Watson (Figure 9.8) and in cooperation with NOAA (National Oceanic and Atmospheric Administration),

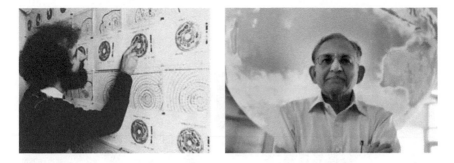

Figure 9.8. (Left) While supporting the International Ozone Trend Panel, Robert Watson examined ozone observations by the TOMS instrument. (Right) P. K. Bhartia produced and released the first ozone observation of the Antarctic ozone hole. Source: https://climate.nasa.gov/news/781/discovering-the-ozone-hole-qa-with-pawan-bhartia/.

FAA (Federal Aviation Administration), WMO (World Meteorological Organization), and UNEP (United Nations Environment Programme), formed in October 1986 an "International Ozone Trend Panel" to reevaluate satellite and ground based observations and to establish with more certainty if the ozone abundance had decreased in the last decade and if this decrease was consistent with our understanding of the effects of natural variability and human activities. The Panel conducted a detailed analysis of all available ozone data and concluded from the information provided by the ground-based Dobson instruments (considered to be the most accurate) that the ozone column between 30° and 64°N had decreased by approximately 1.7 to 3.0% since 1970 and, as suggested by Harris, that the erosion was most pronounced in winter. The Panel also indicated that the satellite data had been affected by the degradation of the diffuser plate of the instrument used to make solar observations, and therefore overestimated the ozone decline. In 1989, the TOMS team found a way to correct for the instrument drift and showed that the trend derived by the space instruments (after correction) was similar to the trend deduced from the measurements by the ground-based Dobson spectrophotometers.

Three Hypotheses to Explain the Formation of the Ozone Hole

The announcement that a hole was forming in the ozone layer attracted the attention of the press and of the broad public. All kind of hypotheses were formulated to explain this phenomenon, and not all of them were rational.

Figure 9.9. Tabloid *Sun* in its December 16, 1986, edition suggests that the ozone hole is produced by the presence of aliens at the South Pole who have been spotted and are burning ozone.

The populist press even suggested that aliens had been found at the South Pole and that *ET*s were burning the Earth's ozone (Figure 9.9). These explanations, although sensational, did not convinced the scientists who rather decided to investigate how dynamical and chemical processes could lead in just a few weeks to a nearly total depletion of ozone in the cold Antarctic lower stratosphere during the month of September.

Three hypotheses were soon formulated to explain the formation of this mysterious ozone hole appearing at the end of the Antarctic winter. The first one came from a scientist then at Clarkson University in Potsdam, NY, K. K. Tung, who stated in 1986 that the ozone hole was probably produced by a reversal of the atmospheric circulation, normally descending over Antarctica, and now probably ascending. This would result in an influx of air from the lower stratospheric and tropospheric layers, both poor in ozone. Although supported by the then Director of NOAA's Geophysical Fluid Dynamics Laboratory, Jerry Mahlman, this theoretical hypothesis was rather quickly dismissed as no anomalies were observed in the vertical distribution of other atmospheric gases that were also expected to be affected by the disruption of the atmospheric circulation.

Linwood B. Callis and Murali Natarajan, two scientists at NASA's Langley Research Center in Virginia, offered a second explanation the same year. These scientists linked the formation of the ozone hole to the 11-year cycle of solar activity. They indicated that, when the Sun was active in the early 1980s, high amounts of nitrogen oxides were produced at an altitude of about 100 km by energetic particles ejected by the active Sun. Within the polar vortex, the nitrogen oxides were probably transported to lower altitudes by the atmospheric circulation, and destroyed stratospheric ozone in the lower stratosphere by the catalytic reactions identified 15 years earlier by Paul Crutzen. The hypothesis of Callis and Natarajan was also rejected because the nitrogen oxides should have destroyed the ozone well above 26 km of altitude—which was not the case—and because the ozone hole should have disappeared by the end of the 1980s after the Sun had regained its weak state of activity.

A third hypothesis was formulated and published on June 19, 1986, in the journal *Nature* by Susan Solomon, a chemist at the NOAA Aeronomy Laboratory in Boulder (Figure 9.10), Rolando R. Garcia, a specialist of the upper atmosphere at NCAR, also at Boulder, F. Sherwood Rowland of the University of California at Irvine, who had highlighted in the 1970s the role

Figure 9.10. Susan Solomon showed that chlorofluorocarbons are responsible for the formation of the ozone hole in Antarctica. Courtesy Susan Solomon, Massachusetts Institute of Technology.

of CFCs in ozone depletion, and Donald J. Wuebbles, then at the Lawrence Livermore National Laboratory in California. Susan Solomon, who completed her doctoral thesis under the supervision of Paul Crutzen and Harold Johnston, was convinced that industrial chlorine, whose stratospheric concentration had increased steadily over the last few decades, was responsible for the formation of the ozone hole. In the stratospheric region where ozone depletion occurs, however, most of the chlorine released by the CFCs is in principle in the form of two relatively inert "reservoirs" without any action on ozone: hydrogen chloride (HCl) and chlorine nitrate (ClONO$_2$). Solomon concluded that, to form reactive chlorine capable of destroying ozone, these two reservoirs had to be converted to chlorine atoms, for example, by reacting together. The process is referred to as *chlorine activation*. This idea, however, did not seem very promising because through laboratory investigations, Rowland had determined that the gas phase HCl + ClONO$_2$ reaction is very slow and, as such, cannot play any important role.

Solomon then recalled that thin clouds, composed of fine ice particles, are formed in the polar stratosphere during the winter when temperatures are sufficiently low. The presence of these clouds had already been observed in 1890 by the Norwegian physicist and mathematician Fredrik Carl Mülertz Størmer (known as Carl Størmer, 1874–1957), who, in a paper written in German, called them *Perlmutterwolken* (in English: mother of pearl clouds). However, the presence of such clouds was found to be rare in the boreal regions, and Størmer had to wait until the end of 1929 (Figure 9.11) for the

Figure 9.11. Polar stratospheric clouds observed near Oslo, Norway, on January 13, 1929. Reproduced Chapman, S., *Nature* 129 (1932), 497–99.

reappearance of these stratospheric clouds. Sydney Chapman who was interested in these clouds wrote in 1932 as a premonitory statement in the journal *Nature*:

> The physical implication of the presence of these clouds at considerable height in the stratosphere may remain obscure, but possibly they will play a significant part, along with such other remarkable facts as those of the ozone relations with surface weather.

Using the spaceborne instrument called SAGE (Stratospheric Aerosol and Gas Experiment), Patrick McCormick at NASA Langley Research Center showed that, indeed stratospheric clouds are episodic in the northern hemisphere, but are almost continuously present in Antarctica during the winter, precisely between the altitudes of 16 and 26 km where the temperature drops below $-80°C$ during the polar night. These clouds are composed of droplets or crystals formed from a mixture of water, sulfuric acid, and nitric acid. Solomon then suggested that the two chlorine reservoirs, HCl and $CONO_2$, could be transformed by so-called *heterogeneous reactions*[3] on the surface of the cloud particles, and thus produce the reactive chlorine atoms capable of destroying ozone after the return of the Sun in the early austral spring. Clouds not only "activate" chlorine but they also "denitrify"[4] the low polar stratosphere since they convert nitrogen oxides into nitric acid, and thus prevent the reformation[5] of the relatively stable $ClONO_2$ reservoir.

Explanation of the Causes of the Ozone Hole

The discovery of the ozone hole had a strong resonance in political circles. The scientific community was challenged by national governments and international organizations to explain the discovery made by the team of the British Antarctic Survey. Already in September 1986, a first ground-based field campaign was organized by NOAA and NASA under the name

---

3. Reactions between the gases present in the air and the chemical species dissolved in the liquid or solid particles, or on the surface of the latter.

4. Nitrogen compounds and in particular nitric acid are rapidly dissolved in cloud particles and removed from the stratosphere when these particles are transported to the troposphere by gravitation.

5. By reaction between nitrogen dioxide NO_2 and chlorine oxide ClO.

of National Ozone Expedition (NOZE) to test the various hypotheses that had been proposed to explain the formation of the ozone hole. The leadership of the first mission was entrusted to Susan Solomon (Figure 9.11) who traveled with a small group of scientists and engineers to the scientific base of McMurdo in Antarctica for several weeks. During the field campaign, chlorine monoxide was observed by the microwave spectrometer developed by Robert de Zafra from the State University of New York at Sony Brook, and the column abundance of the OClO molecule was measured by the NOAA group (Figure 9.12). Measurements of the vertical ozone profile were made above the Amundsen-Scott base at the South Pole by David Hoffman from the University of Wyoming, using balloon-borne ozone sondes. They showed that ozone had virtually disappeared in an altitude range of 16 to 26 km. All these measurements provided the first experimental evidence of abnormally high levels of reactive chlorine and related ozone depletion in the lower atmosphere inside the polar vortex. This led Solomon to state during a press conference that she gave remotely before leaving McMurdo that the chlorofluorocarbons hypothesis was supported by the results of the field campaign.

Figure 9.12. The first Antarctic field campaign at the polar station of McMurdo led by Susan Solomon (left). Using balloon-borne (right) and ground-based instrumentation, the NOAA group working together with university groups observed elevated amounts of chlorine in the stratosphere and concluded that the ozone hole was most probably produced by reactive chlorine. Courtesy: Susan Solomon, Massachusetts Institute of Technology.

This statement generated some controversy: in particular among the meteo-rologists who claimed that the dynamical explanation had not been proven to be incorrect. When she returns from Antarctica, Solomon said

> One of the stunning experiences of my life in science was to witness the ozone levels drop at McMurdo during September 1986. Ozone fell from about 300 Dobson units when we arrived in Antarctica in late August 1986 to less than 200 Dobson units by late September, and the disappearance of a third of the total ozone supported the observations of others regarding the veracity of the ozone hole and its seasonality.

A second campaign to further understand the processes involved in the formation of the ozone hole called the Airborne Antarctic Ozone Experiment (AAOE) took place in August and September 1987. An additional and very important platform was made available by NASA: a former U-2 spy plane converted into a research aircraft and now called the ER-2 (Figure 9.13). Several instruments on board the aircraft, based at the airport

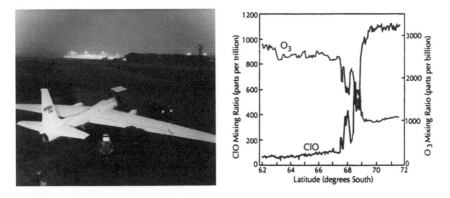

Figure 9.13. (Left) The NASA ER-2 aircraft at the airport of Puenta Arenas, Chile, during the AAOE experiment in September 1987. Photo Bruce Gary, NOAA Earth System Research Laboratory, https://www.esrl.noaa.gov/csd/projects/aaoe/narratives/gary.html. (Right) Simultaneous measurements of the ClO and ozone abundance from 62°S to 72°S on September 16, 1987, made from the ER-2 aircraft flying at 20 km altitude. As the airplane penetrates inside the polar vortex at about 69°S, the concentration of reactive ClO increases dramatically and at the same time the concentration of ozone decreases substantially. These measurements highlight the anticorrelation of ClO and ozone during the Antarctic springtime: where the ClO concentration is high, the ozone concentration is low. Source: Anderson, J. G. et al., *J. Geophys. Res.* 94 (1989), 11465–479.

of Punta Arenas in Chile, directly measured ozone and chlorine oxide (ClO) at an altitude of 20 km. The flights were performed in August and September 1987. The data were unambiguous: in the Antarctic lower stratosphere, the amount of chlorine oxide inside the polar vortex was 100 times higher than outside the vortex. And where chlorine had been activated, ozone had practically disappeared (Figure 9.13). The hypothesis put forward by Susan Solomon and her colleagues became increasingly credible.

Less than two months after the completion of the campaign, a group of prominent atmospheric chemists and meteorologists gathered in Berlin, Germany, at a workshop organized by an organization promoting international scientific exchanges and called the "Dahlem Conferences."[6] Among the participants were Sherry Rowland, Joe Farman, Richard Stolarski, Ralph Cicerone, Paul Crutzen, James Lovelock, Patrick McCormick, Jerry Mahlman, Ivar Isaksen, and Robert Watson. During several days, a working group attempted to put together the pieces of the "ozone hole puzzle." They examined the data collected during the two field campaigns and synthesized all information available to provide a clear picture of the processes involved. The relative role of atmospheric dynamics and chemistry was still a matter of debate, and the scientists were therefore careful not to draw to any rapid conclusion that was not fully supported by the observational data. They wrote in their report:

> Although the preliminary findings from the more recent field experiments have provided much insight about the mechanisms involved in the Antarctic ozone depletion, it remains difficult at this point to provide a global and definitive explanation of the causes of this dramatic phenomenon [...].

At the same time, however, they clearly stated that

> The formation of the "ozone hole" arises from efficient chemical mechanisms. [...] However, dynamics appears to play an important role in setting up the cold PSC-containing vortex, which is required for heterogeneous processes to occur.

The chlorine theory proposed by Susan Solomon and her colleagues was therefore confirmed, but the lack of quantitative information on the

---

6. The Dahlem Conferences, initiated in 1974 to stimulate international cooperation through the organization of interdisciplinary workshops, was named after a district of Berlin that has a long-standing tradition in the sciences and the arts.

importance of heterogeneous reactions was also highlighted. The need for investigating the microphysical processes involved in the formation of PSC particles and to quantify the rates of the reaction occurring at the surface of these particles was identified as a research priority. This topic became therefore the focus of laboratory investigations during the following years. Further, atmospheric observations showed that some of these stratospheric particles in the liquid phase contain a mixture of water, nitric acid, and sulfuric acid. Frozen water particles are present under very cold temperatures.

Solomon continued to investigate the processes involved in the formation of the ozone hole and became interested in the relations between atmospheric chemistry and climate. While at the NOAA Aeronomy Laboratory in Boulder, she chaired the group that produced in 2007 one of the three reports produced for the Fourth Assessment by the Intergovernmental Panel on Climate Change (IPCC). IPCC was awarded the Nobel Peace Prize in 2007. Solomon, a member of the American and French Academies of Sciences, joined the Massachusetts Institute of Technology in 2011.

After the 1986 paper by Solomon was published by *Nature*, the chemistry community attempted to respond to an important question: what are the exact chemical mechanisms involving activated chlorine that lead to the destruction of lower stratospheric ozone in just a few weeks in early spring? The catalytic cycle proposed by Stolarski and Cicerone in 1984 could not provide the answer because it implies the existence of a reaction between chlorine monoxide (ClO) and the oxygen atom (O). Oxygen in atomic form is very scarce in the lower stratosphere and therefore this reaction occurs at a very low rate. The solution was provided in 1987 by Luisa Tan Molina and Mario Molina (Figure 9.14), who showed through chemical kinetics studies carried out in their MIT laboratory that chlorine monoxide can be converted back into a chlorine atom Cl by a series of reactions that involve the formation of the dimer ClO-ClO (also called Cl_2O_2) without requiring the presence of an oxygen atom.[7] When all of these previously ignored processes were introduced in mathematical models, the formation of the ozone hole over the Antarctic continent during the southern spring could be simulated as soon as the calculated temperature was sufficiently low to allow polar stratospheric clouds to form.

---

7. The main catalytic cycle of ozone destruction in the Antarctic region consists of the following reactions: $2(Cl + O_3 \rightarrow ClO + O_2)$, $ClO + ClO + M \rightarrow Cl_2O_2 + M$, and $Cl_2O_2 + hv \rightarrow 2Cl + O_2$. The symbol M represents an atmospheric molecule (N_2 or O_2).

Figure 9.14 Luisa Tan Molina (upper panel) and Mario Molina (lower panel) who identified the catalytic cycle responsible for the rapid destruction of ozone by chlorine in the springtime Antarctic. Source: Beijing Normal University, Faculty of Geographical Science, https://geo.bnu.edu.cn/xwzx/75416.html and American Academy of Achievement, https://www.achievement.org/achiever/mario-j-molina-ph-d/.

Despite this significant progress, which clearly showed the detrimental effects of the chlorofluorocarbons, some of these findings were questioned by pro-industry lobbies and a few other personalities, mainly in Europe and in the United States (Box 9.2). One of the most outspoken "deniers" in North America was S. Fred Singer, the President of the Science & Environmental Policy Project. Singer had played a considerable role in the past: he

Box 9.2. Merchants of Doubts

The research produced by a scientist is valuable only if his/her results can be verified and reproduced by other scientists. Papers disseminating research results, generally by scientific journals, are therefore peer-reviewed prior to publication. The purpose is to examine if the hypotheses adopted by the author(s) are plausible, if previous work is adequately cited, if errors in the reasoning or in the calculations are made and if the conclusions expressed in the paper are logical and reflect the reported findings. The evaluation of the submitted articles is carried out by experts appointed by the journals and unknown to the authors. These authors have an opportunity to respond to the reviewers' comments and modify their papers if they wish so. A significant proportion of the articles are rejected because they are considered to bear too little innovation or to provide unreliable results. Other articles that make a significant contribution are accepted and published. This practice is important because it contributes to the quality of the science and the objectivity of the research. All studies devoted to the effects of human activity on the ozone layer or on climate are the subject of particular scrutiny because of their significant impact for society.

Despite the consensus that has developed within the scientific community on the issues related to atmospheric ozone, sometimes after long technical debates and professional controversies, a handful of scientists, often with prominent past records, have tried in the 1980's and 1990's to cast doubt on the consensus by asserting that the research lacked objectivity and that the conclusions were incorrect and biased. Hence, they should not be taken into consideration by decision-makers. Several of these *deniers*, sometimes called *contrarians*, belong to the generation of scientists that was strongly influenced by the economic and military role played at that time by the two superpowers. Environmental studies started to show that the prevailing economic growth of the golden sixties was unsustainable. As defenders of the free enterprise, this group challenged the scientific conclusions on most environmental issues of the time, including ozone depletion, acid rain, nuclear winter, climate change; they even questioned the fact that asbestos and tobacco (passive smoking) have detrimental effects on human health.

In the United States, Earth observation physicist Siegfried (Fred) Singer, solid-state physicist, Fred Seitz (1911–2008) (who was associated with the development of the atomic bomb during the Second World War), and physicist William Aaron Nierenberg (1919–2000) (also associated with the Manhattan Project and former director of the Scripps Institution of Oceanography) played a major role in attacking the findings provided by environmental scientists. Their actions were supported by a number of foundations including the George C. Marshall and the Heartland Institutes, two conservative think tanks with focus on science and policy, and the Global Climate Coalition, an international lobbyist group of businesses that challenged the science behind global warming. Their work was encouraged and even financed by several multi-national corporations.

One of the most virulent actors against the results of ozone science was Fred Singer. The latter denounced in several newspapers including the *Wall Street Journal* and the *National Review* what he called "the ozone scare". He first attacked those who had brought to light in the early 1980s the possible erosion of the ozone layer by the supersonic aircraft. He then turned his attention to the problem of chlorofluorocarbons, believing that these compounds had no discernible effect on ozone. For him, the ozone hole was a purely natural phenomenon, probably caused by the chlorine emitted in the atmosphere during volcanic eruptions, a phenomenon, he said, that had been already observed by Dobson in the 1930s. The ozone hole was therefore expected to fill quickly. In any case, it was superfluous to reduce the industrial production of chlorofluorocarbons since, according to Singer, these were products harmless to the environment.

No rational scientific explanations based on verified experimental facts were ever proposed by any of the *deniers*, and specifically no alternative theory ever passed the peer-review process imposed by scientific journals. The arguments provided by Singer and his colleagues were simplistic statements that ignored many of the established facts. They were always presented in mainstream newspapers, politically motivated books, or at parliamentary hearings. Another prominent contrarian, Dixy Lee Ray (1914–1994), a marine biologist, former president of the Atomic Energy Commission and former governor of Washington State, published a book entitled *Trashing the Planet*. where she questioned the science related to ozone depletion, acid rain and climate change. Regelio Maduro and Ralf Schauerhammer produced a popular volume under the title *The Holes in the Ozone Scare: The Scientific*

Evidence That the Sky Isn't Falling. The popular radio broadcasts by the far-right commentator, Rush Limbaugh, were designed to amplify incorrect scientific statements produced by the *deniers* with a political rather than a scientific objective. Limbaugh in his book entitled *The Way Things Ought to Be* insists that the theory of ozone depletion by CFCs is a hoax: "balderdash" and "poppycock".

In their book, *The Merchants of Doubt. How a Handful of Scientists Obscured the Truth on Issues from Tobacco Smoke to Global Warming*, Naomi Oreskes and Erik Conway very clearly explain the method used by those who constantly question the scientific consensus. They write

> One of the strategies to cast doubt on ozone depletion was the development of a counter narrative that portrayed the depletion of ozone as a natural variation, cynically exploited by a corrupt, selfish scientific community. and extremist, to get more money for his research.

Oreskes and Conway highlight the motivations behind this strategy. *Deniers*, supported by politically engaged think tanks and in some cases by industrial and banking groups, have consistently opposed measures taken to protect the environment, including reduce acid rain, protect climate and limit tobacco consumption. The authors conclude that the motivation of money is essentially ideological:

> Orwell understood that those in power will always seek to control history, because whoever controls the past controls the present. So, our Cold Warriors – Fred Seitz and Fred Singer [...] and later Dixy Lee Ray, who had dedicated their lives to fighting Soviet communism, joined forces with the self-appointed defenders of the free market to blame the messenger, to undermine science, to deny the truth, and to market doubt. People who began their careers as fact finders, ended them as fact fighters. Evidently accepting that their ends justified their means, they embraced the tactics of their enemy, the very things they had hated in Soviet communism for: its lies, its deceit, its denial of the very realities it had created.

And they conclude,

> Why did this group of Cold Warriors turn against the very science to which they had previously dedicated their lives? Because they felt [...] they were

working to "secure the blessings of liberty". If science was being used against those blessings – in ways that challenged the freedom of enterprise – then they would fight it as they would fight any enemy. For indeed, science was starting to show that certain kinds of liberties are not sustainable – like the liberty to pollute. Science was showing that Isaiah Berlin was right: liberty for wolves does indeed mean death to lambs.

Deniers were also vocal in other countries, particularly with regard to the origin and evolution of global warming. In France, for example, the former Minister of Education and Research, and world-class geochemist, Claude Allègre, a recipient of the Crawford prize and auteur of the book *L'imposture climatique* (Climate Imposture), qualified ecology as nothing else than a protest by weak scientists who have identified a lucrative business. The motivation of Allègre and of another activist *denier* in France, volcanologist Haroun Tazieff (1914–1998), was probably less a defense of the free enterprise than a rejection of the concept of scientific consensus that appeared to be so authoritative that it let a very small space for critical discussions.

What emerged from this intense and often unpleasant controversy is that a dialogue between scientists and *deniers* was impossible because the debates were less about the scientific issues than about societal values. The debate became therefore essentially ideological.

was among the initiators of the International geophysical Year and more recently had contributed to the development of advanced space instruments to measure ozone. In a testimony before a subcommittee of the House of Representatives, he said in 1995:

In fact, the history of the CFC-ozone depletion issue is rife with selective use of data, faulty application of statistics, disregard of contrary evidence, and other scientific distortions. [...] The hypothesis that CFCs deplete ozone is still just that: a hypothesis. The theory did not predict the Antarctic ozone hole and cannot predict what will happen globally. There is no firm evidence as yet for long-term depletion of global ozone.

Are halocarbons, like CFCs, really the major culprit when it comes to ozone damage? Let me give you a different scenario, but one which is also scientific plausible. [...] It is quite possible [...] that the controlling factor in the creation of an ozone hole in the Antarctic—and potentially elsewhere—are methane and carbon dioxide rather the growth in CFCs.

In France, the famous volcanologist Haroun Tazieff, who had been working in Antarctica did not hesitate to say:

> I wonder if CFCs are not accused of destroying the ozone layer for reasons that are more economic than ecological. Because there's a lot of money to be made. But monies invested in gigantic battles against imaginary pollution are no longer available to fight real pollution.

The denialist tactic was to side step the studies published in peer-reviewed scientific journals, and rather to publish newspapers articles and give testimonies before Congress. In fact, no serious scientific counterargument or an alternative theory was ever presented by these "deniers," who seemed to be exclusively motivated by their ideology and their political agenda, and sometimes supported by some specific lobbies. Despite these pressures to the governments, the countries that had signed the Montreal Protocol decided to take additional measures that ultimately led to the phase-out of the CFCs. The industry, which for a long time had attempted to convince decision makers that their products were harmless, accepted this verdict and started to develop alternative products that are now being used in refrigeration systems and other applications. In fact, throughout this period, industry, initially reluctant, quickly engaged in a constructive process of supporting science and, after expressing reservations about Molina and Rowland's assumptions, suggested that it would accept the researchers' verdict. The Montreal Protocol would not have been concluded without the exchange of information between the scientific community, industry, governments, and international institutions.

Despite decisions made in the 1980s and 1990s, the ozone hole has not yet disappeared. The lifetime of the CFCs in the stratosphere is several decades and it will therefore be necessary to wait until the second half of the twenty-first century to find conditions similar to those prevailing in the 1970s before the ozone hole appeared. Although the first signs of a "recovery" of the ozone layer have emerged in recent years, "recovery" will take a few more decades (Figure 9.15) and is likely to occur in a stratosphere that will become gradually colder in response to continued CO_2 emissions. It is therefore likely that the stratosphere will never return to a state that is exactly the same as in the 1960s.

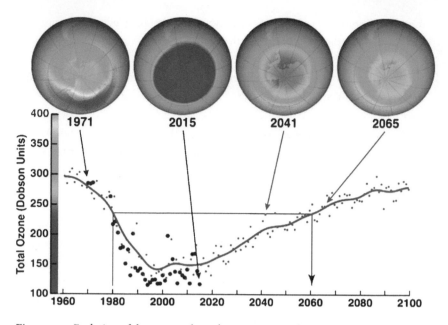

Figure 9.15. Evolution of the ozone column between 1960 and 2100 in Antarctica. The lower part of the figure shows the values of the total vertical ozone quantity observed up to 2015 (black dots). The red line shows the values calculated by a mathematical model for the past and projected values for future years. The amount of ozone is expected to return to its 1980 level only by 2060. The upper part of the figure shows satellite observations of the ozone hole in 1971 and 2015 and a similar representation of the future evolution of the ozone hole in 2041 and 2065, as simulated by a global atmospheric model. Credit: NASA Goddard Space Flight Center.

An Ozone Hole in the Arctic?

Finally, it remained to be understood why the ozone hole observed in Antarctica did not occur in the Arctic as well, since CFC emissions were taking place mainly in the northern hemisphere. As stated in Chapter 6, the atmospheric dynamics of both hemispheres is far from being identical. Few planetary waves develop in the southern hemisphere and the polar vortex surrounding Antarctica is therefore very stable[8]; it isolates a very

---

8. Occasionally, for example, in 2002 and 2019, large planetary waves developed in the southern hemisphere, which lead to an unusual warming of the Antarctic stratosphere and hence to a small ozone hole. In 2019, the peak extent of the ozone hole was 16.3 million km² to be compared with a value of 21 million km² in more typical years.

cold polar region from any external influences during the winter. The austral polar temperature is sufficiently low to allow stratospheric clouds to form and to remain present during most of the winter. On the other hand, large-scale atmospheric waves disrupt much of the northern hemisphere, which destabilizes the Arctic's polar vortex and allows thermal energy and ozone to be transported to the North Pole. As a result, the temperature of the Arctic stratosphere is on average 10°C higher than that of the Antarctic stratosphere, and, as Størmer noted earlier, clouds are formed, but only occasionally. The conditions for abrupt ozone depletion in the lower boreal stratosphere are not met most of the time. Only during a few winters characterized by an unusually stable polar vortex and low polar temperatures in the lower stratosphere, a considerable reduction in ozone is observed, especially over Scandinavia, Canada, and northern Russia. It remains to be determined whether the cooling of the stratosphere, which results from the human-induced increase in atmospheric CO_2, will produce more frequent ozone holes in the Arctic before the CFCs have been completely removed from the stratosphere. Nobody is safe from a future nasty surprise.

CHAPTER TEN

Ozone in the Troposphere

T hanks to the work of Jean Auguste Houzeau in Rouen, France, it is known since 1858 that ozone is a permanent constituent of the lower layers of the atmosphere. More recent observations have shown that this gas is often most abundant in the vicinity of populated areas, mainly during the summer and when surrounding air masses are stagnant. Ozone is a powerful oxidant that causes health problems when its concentration is high: it irritates the throat and eyes, promotes the development of bronchitis and bronchopneumonia, produces asthma attacks, causes other lung diseases, and weakens our immune system. It also promotes the onset of heart disease and stroke. Besides these health impacts, it reduces the productivity of crops and hence the production of food.

Air Pollution

Acute pollution episodes that occur during summertime in large metropolitan areas, referred to as "Los Angeles-type smog" are associated with high concentrations of ozone. During wintertime, primarily in regions affected by intense coal burning, the presence of fine solid or liquid particles that are either emitted at the surface or formed in the atmosphere by nucleation[1] or condensation processes, produce a "London-type smog." The chemical composition of fine particles in the atmosphere, also known as aerosols, is complex and varies with the location where they are formed. In addition to sulfuric acid produced by the combustion of sulfur-rich coal, aerosols may contain nitric acid or various organic acidic compounds. They are at the origin of acid rain that destroys vegetation and is notably responsible for the *Waldsterben*[2] that was observed in the 1960s, particularly in the Nordic European countries, in Canada, and still today in China and India. Carbon particles resulting from agricultural, industrial, and domestic combustion processes, dust from deserts, and sea salt released from the oceans are also present in the atmosphere. Large amounts of particles are also released by wildfires and volcanic eruptions. Fine particles are a source of pulmonary and cardiac diseases, and probably promote the development of lung cancer. The World Health Organization (WHO) estimates that today nearly eight million people worldwide die prematurely each year from indoor and outdoor air pollution (particularly particulate matter). Developing countries such as China and India, and now also African countries, are particularly affected by urban air pollution, and very intense pollution episodes are observed in densely populated areas. Some of these countries, which are aware of the problem, are taking measures to reduce air emissions of pollutants, but with insufficient results.

The problem of air pollution is not new. Already in 1306, King Edward decided to ban the burning of coal, which produced considerable quantities of sulfide fumes and was a major source of air pollution in London. With the Industrial Revolution, pollution episodes became more frequent, especially during the winter when temperature inversions[3] occurred that stabilized air masses in contact with the surface. Some of these dramatic episodes will be remembered: that of the Meuse valley in Belgium, one of the most industrialized regions in Europe in the early part of the twentieth century, where a persistent fog formed from December 1–5, 1930. This episode

---

1. Nucleation: Initial process that leads to the formation of crystals from a liquid or vapor.

2. Waldsterben: Conditions in which trees die, for example, under the effect of acid precipitation.

3. In most cases, the temperature of the troposphere decreases with altitude. However, when the ground is cold during the winter, the temperature can increase with altitude in the lower layers of the atmosphere (boundary layer), which makes the air masses very stable (stagnant air) near the ground.

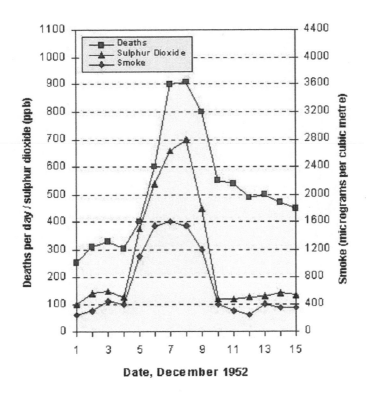

Figure 10.1. Concentration of sulfur dioxide (ppb) and of smoke (μg/m³) and the related number of deaths per day experienced in London during the tragic smog episode in December 1952. Source: http://www.air-quality.org.uk/03.php.

occurred between the cities of Huy and Liège, where 6,000 residents suddenly suffered from respiratory problems and at least 60 of them died. Another similar event happened in Donora, Pennsylvania, during October 1948. Forty-two percent of the population was affected and 20 people lost their lives. The persistent fog in London during December 1952 (Figures 10.1 and 10.2) caused 12,000 deaths, even though the Prime Minister at the time, Winston Churchill, minimized the case by considering that it was a banal meteorological event that would quickly end.

Ozone Pollution

After the Second World War II, measures were taken to reduce emissions of sulfur compounds from industrial complexes with satisfactory results in Europe and in several North American cities. Yet pollution persists,

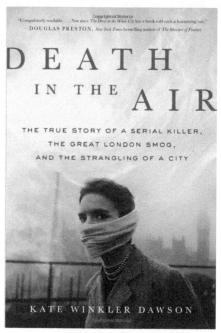

Figure 10.2. (Upper panel) In December 1952 during an acute smog event, London was trapped in a deadly cloud of fog and pollution for five days. A police officer on duty uses flares to guide the traffic. Source: (https://www.theguardian.com/commentisfree/2017/dec/05/smog-day-warning-18000-die-london-great-smog-1952-air-pollution). (Lower panel) The book of Kate Winkler Dawson is a historical narrative of this environmental disaster known as a "serial killer" in the iconic city of London. Source: https://www.katewinklerdawson.com/.

especially during the summer months around certain urban areas such as Los Angeles, where the population was experiencing eye irritation and respiratory problems. Analyses of the composition of the air showed that the *smog* events in southern California were not dominated by the presence of fine particles as was the case, for example, in London, but were associated with high concentrations of ozone. Such episodes occurred mainly during the summer with ozone concentrations reaching occasionally more than 500 ppbv.

As stated in the previous sections, the first step required for the formation of ozone is the production of oxygen atoms (O). Chapman has shown in 1929 that the photodissociation of molecular oxygen (O_2) by short-wave radiation is the source of these oxygen atoms in the stratosphere. This particular process, however, cannot be invoked at lower altitudes in the troposphere because the ultraviolet radiation (at wavelengths less than 242 nm) that is sufficiently energetic to break the oxygen molecule into two oxygen atoms is totally absorbed above 20 km altitude. Therefore, it cannot initiate the formation of the urban ozone episodes. Another chemical mechanism needs therefore to be invoked.

A first explanation of the presence of ozone in the troposphere was proposed several decades ago: ozone-rich air masses could be transported downward by intrusions of air from the stratosphere, specifically along the jet streams during the passage of weather disturbances. These intrusions of ozone are indeed observed, but they cannot account for the presence of high concentrations of ozone within standing air masses near the ground in urban areas. There must therefore be a local chemical source of ozone, probably associated with human activity. However, until the middle of the twentieth century, this missing mechanism was unknown. It is the Dutch biologist and chemist Arie Jan Haagen-Smit who provided the missing explanation in 1950.

Haagen-Smit (1900–1977; Figure 10.3) was a graduate of the University of Utrecht with a PhD on hydrocarbons emitted by certain plants, specifically the so-called sesquiterpenes. After a brief visit at Harvard University in 1936, Haagen-Smit joined the California Institute of Technology in Pasadena, where he became a professor of biogeochemistry in 1940. There, he became aware of the effects of the California *smog* on plant growth and on the health of the region's inhabitants. He showed that the high concentrations of ozone observed in Los Angeles, probably responsible for the "new plant diseases" that the farmers had noticed in the region, resulted from the combined presence of hydrocarbons and nitrogen oxides emitted by automobiles

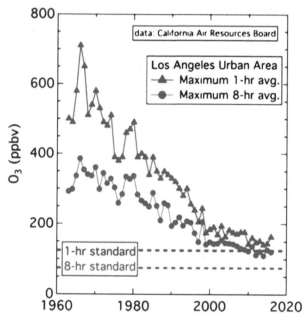

Figure 10.3. (Upper panel) Arie Jan Haagen-Smit who discovered the mechanisms of ozone formation in urban and industrial areas. Credit: Archives, California Institute of Technology. (Lower panel) Changes (decrease) in the concentration of ozone (maximum 1- and 8-hour averages) in the Los Angeles urban area between 1962 and 2017 as a result of actions taken to reduce emissions of pollutants that are a source of ozone. Source: https://eos.org/features/urbanization-air-pollution-now.

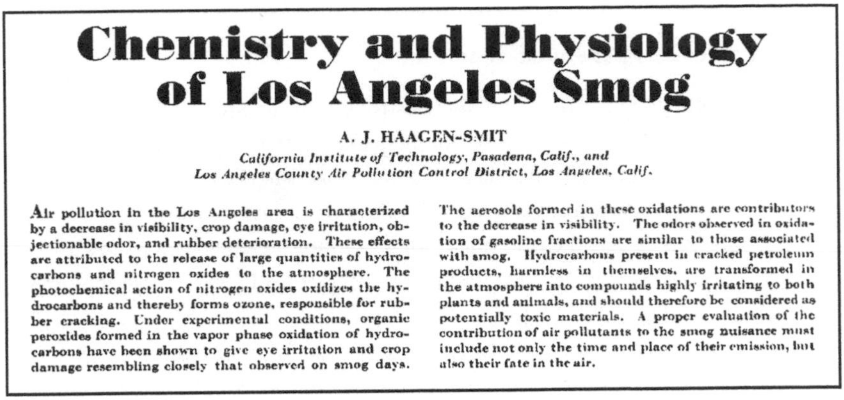

Chemistry and Physiology
of Los Angeles Smog

A. J. HAAGEN-SMIT

California Institute of Technology, Pasadena, Calif., and
Los Angeles County Air Pollution Control District, Los Angeles, Calif.

Air pollution in the Los Angeles area is characterized by a decrease in visibility, crop damage, eye irritation, objectionable odor, and rubber deterioration. These effects are attributed to the release of large quantities of hydrocarbons and nitrogen oxides to the atmosphere. The photochemical action of nitrogen oxides oxidizes the hydrocarbons and thereby forms ozone, responsible for rubber cracking. Under experimental conditions, organic peroxides formed in the vapor phase oxidation of hydrocarbons have been shown to give eye irritation and crop damage resembling closely that observed on smog days.

The aerosols formed in these oxidations are contributors to the decrease in visibility. The odors observed in oxidation of gasoline fractions are similar to those associated with smog. Hydrocarbons present in cracked petroleum products, harmless in themselves, are transformed in the atmosphere into compounds highly irritating to both plants and animals, and should therefore be considered as potentially toxic materials. A proper evaluation of the contribution of air pollutants to the smog nuisance must include not only the time and place of their emission, but also their fate in the air.

Figure 10.4. Paper published by Arie Haagen-Smit in 1952 attributing the formation of ozone to the release of hydrocarbons and nitrogen oxides to the atmosphere. Reproduced from Haagen-Smit, A. J., *Ind. Eng. Chem.* 44 (1952), 1342–46.

and industrial facilities (Figure 10.4). This suggestion was first strongly contested by representatives of the petroleum industry, and in particular by some members of the Stanford Research Institute (SRI), an institution funded by the industry. Haagen-Smit's findings were judged by the members of this institute to be illogical and his work irreproducible, but his point of view eventually became authoritative. The role of automobiles in *smog* formation became gradually accepted so that a first emission control systems for vehicles was established in the early 1960s. Since then, anti-pollution standards have multiplied. Car engines have gradually been adapted and industrial emissions reduced. As a result, ozone concentrations in populated areas such as Los Angeles have decreased considerably, although they still rather frequently exceed the authorized values (Figure 10.3), mainly during summertime. The relaxation of anti-pollution measures proposed in 2018 by the federal administration of the United States is expected to have adverse effects on human health in different regions of the country.

The photochemical ozone production mechanism identified by Haagen-Smit (Figure 10.3) is complex (Box 10.1) and is based on a catalytic cycle involving nitric oxide (NO), emitted, for example, by automobiles. Nitric oxide is known to be converted to nitrogen dioxide (NO_2) by reaction with ozone,[4] and the dissociation of NO_2 by solar light produces an oxygen

---

4. $NO + O_3 \rightarrow NO_2 + O_2$.

Box 10.1. Smog chemical reactions

The smog reactions producing and destroying ozone are illustrated here in the simple case where carbon monoxide (CO emitted, e.g., at the surface by automobiles) represents the "fuel" that contributes to the formation chain of ozone. The production of the hydroxyl radical OH is initiated by the photolysis of ozone followed by the reaction of the resulting singlet oxygen atom $O(^1D)$ with water vapor. OH is converted to a peroxy radical HO_2 by reaction with carbon monoxide. In the presence of high atmospheric concentrations of NO (polluted environment), the chain continues by a conversion of HO_2 by NO, producing OH and NO_2. This last species is photolyzed during daytime and constitutes a source of ozone. Under low NO concentrations (unpolluted environment), HO_2 destroys ozone directly. The cycles involved are catalytic for nitrogen oxides and the hydrogenated free radicals. The conversion of NO_2 and OH into nitric acid HNO_3 interrupts these catalytic cycles by removing nitrogen oxides and hydrogen radicals. In more complex cases that play a considerable role in urban and industrial areas, the "fuel" is provided by nonmethane hydrocarbons (RH), which are also oxidized by the OH radical and produce organic peroxy radicals (often noted as RO_2). These react with NO to produce NO_2, which constitutes an additional source of ozone.

Radical initiation

$O_3 + h\nu \rightarrow O(^1D) + O_2$

$O(^1D) + H_2O \rightarrow OH + OH$

Radical termination

$OH + NO_2 + M \rightarrow HNO_3 + M$

Ozone production

$CO + OH + O_2 \rightarrow CO_2 + HO_2$

$HO_2 + NO \rightarrow NO_2 + OH$

$NO_2 + h\nu \rightarrow NO + O$

$O + O_2 + M \rightarrow O_3 + M$

----

net: $CO + 2\,O_2 \rightarrow O_3 + CO_2$

Ozone destruction

$CO + OH + O_2 \rightarrow CO_2 + HO_2$

$HO_2 + O_3 \rightarrow OH + 2\,O_2$

----

net: $CO + O_3 \rightarrow O_2 + CO_2$

atom (O) while reforming a nitric oxide molecule NO.[5] The oxygen atom rapidly combines with an oxygen molecule (O_2) to form an ozone molecule (O_3).[6] The solar radiation that penetrates into the lowest layers

---

5. $NO_2 + h\nu \rightarrow NO + O$.

6. $O + O_2 + M \rightarrow O_3 + M$, where M represents an inert molecule of nitrogen or oxygen.

of the atmosphere is sufficiently energetic to dissociate NO_2. Considered together, these photochemical reactions constitute a catalytic cycle that can be repeated many times. Each cycle consumes and produces one ozone molecule, so that the net effect of the cycle is null in terms of the ozone budget. What Haagen-Smit showed is that the NO to NO_2 conversion can happen not only by a reaction with ozone, but also by a reaction with a peroxy radical $(HO_2, CH_3O_2, etc.)$[7] that does not consume ozone. In this case, each catalytic cycle results in the production of ozone. Hydroperoxy radicals (HO_2) are formed by reactions between the hydroxyl (OH) radical and methane, other nonmethane hydrocarbons or carbon monoxide (CO).[8] These latter chemical species are emitted at the Earth's surface by biogenic processes (vegetation) and, primarily in urban areas, by industrial and other human activities (e.g., automobile exhaust). They provide the "fuel" needed to produce ozone in high NOx environments. The role of OH in the atmosphere is fundamental because it determines the capacity of the atmosphere to oxidize a large number of pollutants, and hence to clean itself from pollutants. The next question is therefore to identify how the OH radical is formed in the atmosphere? And does this radical affect the atmosphere beyond the very polluted urban or industrial areas?

Global Tropospheric Chemistry

A first response to this last question came in 1971 from the work of Hiram "Chip" Levy, then at the Harvard-Smithsonian Center for Astrophysics. Levy showed that solar radiation at wavelengths around 300 nm, although strongly attenuated while penetrating in the atmosphere, was sufficiently intense to produce the hydroxyl radical (OH) in the lower layers of the atmosphere. In the mechanism that he suggested, the production of OH is initiated by formation of an oxygen atom in its excited electronic state $(O(^1D))$ resulting from the photodissociation of ozone by this radiation. This very reactive $O(^1D)$ atom oxidizes water vapor molecules present in the troposphere to produce the OH radical.[9]

As indicated above, OH reacts easily with different chemical species. It destroys, for example, carbon monoxide in a few months, and the different nonmethane hydrocarbons in a few days or even in a few hours.

---

7. For example: $NO + HO_2 \rightarrow NO_2 + OH$.

8. $CO + OH \rightarrow CO_2 + H$ followed by $H + O_2 + M \rightarrow HO_2 + M$.

9. $O_3 + h\nu \rightarrow O(^1D) + O_2$ followed by $H_2O + O(^1D) \rightarrow 2OH$.

This explains why most hydrocarbons are destroyed close to the surface before they can be transported to the upper layers of the troposphere. The reaction time scale for the reaction of OH with methane is considerably longer, of the order of eight to ten years. With such a slow degradation rate, methane is therefore dispersed and redistributed in the entire atmosphere and contributes to the formation of ozone in the free troposphere.[10] Being produced by a photochemical process involving ozone and water vapor, OH is present in the entire troposphere during daytime with its concentration being larger in the tropics where the amount of incoming solar radiation is highest.

In the early 1970s, Paul Crutzen became interested in the photochemical processes that determine the photochemical production and destruction of ozone in the troposphere at a global scale, particularly in remote regions and not just in urban areas. While working in Oxford, he decided to extend his stratospheric model downwards and add to his adopted chemical scheme the reactions that Haagen-Smit had proposed for urban conditions 20 years earlier. The first presentation of his results during the 1972 Quadrennial Ozone Symposium in Arosa received little attention because, at that time, the scientific community was mostly concerned by the vulnerability of the stratospheric ozone layer to the nitrogen oxides expected to be released by the future supersonic aviation.

In the United States, William Chameides (Figure 10.5) and James Walker were working on the same question at Yale University in Connecticut. In the early 1970s, they developed a comprehensive photochemical theory of tropospheric ozone that served as a basis for many further studies. Chameides, a native of New York City, left Yale in 1974 to work with Ralph Cicerone at the University of Michigan before joining Georgia Tech where he was appointed as a professor. Chameides was elected to the United States Academy of Sciences in 1998.

After he joined NCAR (National Center for Atmospheric Research), Crutzen remained interested in global tropospheric chemistry. He convinced his postdoctoral fellow, Jack Fishman (Figure 10.5), and his PhD student, Susan Solomon, to quantify the global budget of tropospheric ozone. Both early career researchers showed that this budget is significantly affected by the emissions of chemical compounds produced as a result of human

---

10. The free troposphere is the region of the troposphere situated above the boundary layer, the thin atmospheric layer affected by surface processes and extending from the surface to about 1 to 3 km altitude.

Figure 10.5. Four specialists who contributed to our understanding of the photochemical processes affecting tropospheric ozone. (From top to down and from left to right) William Chameides (source: https://nicholas.duke.edu/people/faculty/chameides), Daniel Jacob (source: http://acmg.seas.harvard.edu/education.html), Anne Thompson (source: https://science.gsfc.nasa.gov/sed/bio/anne.m.thompson), and Jack Fishman (source: https://www.nasa.gov/topics/earth/features/bad_ozone.html).

activities. Fishman, who later joined the NASA Research Center in Langley, deduced from the analysis of space observations that, as expected from the theory, during the summer, the highest concentrations of ozone are found in the industrial areas of the northern hemisphere. He also found—this time unexpectedly—that during the months of September and October, large quantities of ozone are produced in the tropical regions of Africa and America and over the Atlantic Ocean. This "ozone bulge" was attributed to the emissions of "ozone precursors" produced by biomass burning and lightning on the continents. This finding led to further experimental and modeling studies, and, for example, Anne Thompson, a Senior Scientist at NASA Goddard Flight Center (Figure 10.5), established in the tropics and subtropics an ozone monitoring program. A number of ozonesonde stations were established to provide sustained information on tropospheric ozone in regions where data were previously severely missing.

The first studies on tropospheric ozone highlighted the complexity of the photochemistry of the lowest layers of the atmosphere. The heterogeneity in the surface emissions and the multiplicity of meteorological and hydrological patterns required new observational and modeling strategies, including

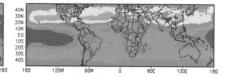

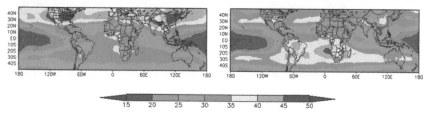

Figure 10.6. Climatology of the tropospheric ozone column (Dobson units) deduced by Jack Fishman and colleagues by subtracting the stratospheric ozone column derived by the SBUV space instrument from the total ozone column measured by the TOMS instrument. The average values of the tropospheric ozone column are shown for northern hemisphere winter (top left), spring (top right), summer (bottom left), and autumn (bottom right). During the four seasons, the ozone amount is very low over the tropical Pacific Ocean. It reaches high values at mid-latitudes in the northern hemisphere during summer (anthropogenic influence). During the months of September to November, high ozone column values associated with tropical fires during the dry season are visible in South America and southern Africa as well as above the Atlantic Ocean, south of the equator. From Fishman et al., *Atmos. Chem. Phys.* 3 (2003), 893–907 (https://www.atmos-chem-phys.net/3/893/2003/).

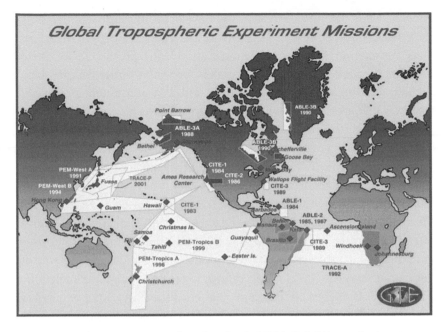

Figure 10.7 Airborne missions conducted under the NASA Global Tropospheric Experiments Program. In red are the airports where the research aircraft made stopovers. The names (acronyms) of the mission are labeled in blue or white. From NASA: www-gte. larc.nasa.gov/gte_map.htm.

three-dimensional approaches. Since the early satellites were unable to detect most chemical species below the tropopause, NASA decided to establish a "Global Tropospheric Experiment (GTE)" (Figure 10.7) under which several airborne field campaigns would be organized in different parts of the world. Joseph (Joe) McNeal, the head of this program, invited prominent scientists to install their instruments on the NASA DC-8 or on other airplanes to measure chemical species along the flight tracks. The analysis of the observations made under different chemical regimes brought a lot of new information which greatly increased knowledge on atmospheric composition and related chemical processes.

The interpretation of observations has been greatly facilitated by the simulations made by mathematical models. These models simulate the chemical transformations and multiscale transport of chemical species emitted or deposited on the surface of the Earth. It is at the Max Planck Institute for Chemistry in Mainz, Germany, that the first global three-dimensional model of tropospheric ozone called MOGUNTIA was initiated in the late 1980s by Peter Zimmermann, one of Paul Crutzen's PhD students. More advanced models were developed in the 1990s including the IMAGES model by Jean-François Müller at the Belgian Institute for Space Aeronomy, the MOZART model at NCAR, and the GEOS-Chem model at Harvard University. The models were first rather rudimentary, but they became increasingly sophisticated and accounted for several processes that are considerably more complex than those encountered in the stratosphere: the formation and dissipation of multiscale meteorological systems, the existence of the hydrological cycle with the effects of clouds and precipitation, and the interactions between the atmosphere and the surface of the Earth. These models, including the Community Atmospheric Model with Chemistry (CAM-Chem, see Figure 10.8), supported by NCAR or the IFS model developed at ECMWF (European Centre for Medium-Range Weather Forecasts) are now used to provide daily global forecasts of ozone and other pollutants. Several of these models are also used in support of field campaigns. The GEOS-Chem model, for example, under the scientific leadership of Daniel Jacob (Figure 10.5) at Harvard University, contributed extensively to the analysis and interpretation of observations made during the NASA GTE program. Local air pollution models that predict the effects of pollution in urban areas (e.g., traffic activity, residential emissions) or the dispersion of plumes produced by industrial facilities are used by environmental services to warn the population of imminent air pollution episodes. Today, integrated models of the Earth system account for the interactions between the atmosphere, the continental biosphere, the ocean, and the cryosphere, with the purpose of projecting long-term changes in the

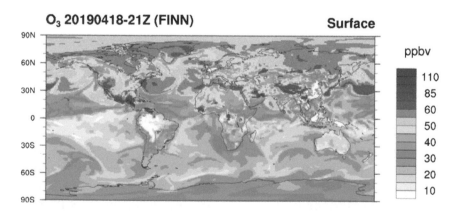

Figure 10.8. Prediction of the global surface ozone mixing ratio (ppbv)] for April 18, 2018 at 21:0 Z. by the global NCAR chemical transport model. In red colors are the hot spots of ozone on this particular day and this particular time of the day. Courtesy: Louisa Emmons, NCAR.

chemical composition of the atmosphere and in climate. The development of data assimilation techniques allows to manage large amounts of data, specifically satellite observations and to combine them with information provided by numerical models.

Laboratory studies, field experiments, and model simulations have provided the information necessary to identify the mechanisms, leading to the production of tropospheric ozone, and thus to establish regulations aimed at limiting the production of this gas by reducing the emissions of ozone precursors (e.g., nitrogen oxides and organic compounds). These regulations have made it possible to reduce air pollution in and near urban areas. However, the effectiveness of the measures depends on the relative emissions intensity of nitrogen oxides and hydrocarbons. In urban centers, the most effective way to reduce ozone in the atmosphere is often to reduce hydrocarbon emissions, while in rural areas or at the outskirts of cities, it is generally preferable to reduce nitrogen oxide emissions. Measures to reduce ozone pollution have been gradually implemented in many urban areas. However, it took the city of Los Angeles and other North American and European urban centers 50 years to solve, at least partially, their air pollution problems. Other cities, especially in Asia, are now taking similar measures, but pollution levels are still often high; progress, however, is underway.

The chemical processes that determine the budget of tropospheric ozone are very different in remote areas of the world. For example, in the polar regions, dramatic ozone depletion is often observed near the surface during spring and is attributed to the explosive release of bromine compounds[11] by surface sea ice and its activation by photochemical processes as solar light becomes available in March (see Box 10.2).

---

11. The rapid ozone destruction above the ice sheet in polar regions results from a catalytic cycle involving a sequence of reactions: First, a reaction of atmospheric HOBr on the surface of the ice leading to the extraction of bromide (Br^-) from the solid ice phase and the release of a molecule of bromine: Br_2: $HOBr + Br^- + H^+ \rightarrow Br_2 + H_2O$. In the presence of sunlight, this heterogeneous reaction is followed by the photolysis of Br_2 releasing two bromine atoms: $Br_2 + h\nu \rightarrow 2Br$, and by the conversion of the Br atoms into bromine oxide BrO: $2(Br + O_3 \rightarrow BrO + O_2)$. Two molecules of HOBr are formed by reaction $2(BrO + HO_2 \rightarrow HOBr + O_2)$. Thus, for each cycle, the concentration of atmospheric bromine compounds increases as additional bromide (Br^-) is extracted from the ice. This amplification mechanism leads to an explosive increase in the concentration of reactive bromine in the lower levels of the atmosphere during spring and hence to rapid ozone destruction.

Box 10.2. Bromine explosion and ozone destruction in the Arctic

The small town of Alert (82.5°N, 62.3°W) located in the Qikiqtaaluk region of the Nunavut Territory in Canada is the northernmost permanently inhabited settlement on the Earth. It is distant by only 817 km from the North Pole. The town has a strategic importance because it hosts a military base of the Canadian Forces that monitors telecommunications emanating from Russia. In addition, a scientific and operational station of Environment Canada provides continuous data on weather and atmospheric chemical composition in this remote region of the world.

In 1986, a team of Canadian scientists led by Jan W. Bottenheim discovered that, near Alert, ozone was almost totally disappearing in the lowest layers of the troposphere during the months of March and April, just after the Sun had returned over the Boreal region at the end of the long winter polar night. Interestingly, a similar phenomenon had already been reported in 1981 by Samuel Oltmans, a scientist working for NOAA in Boulder, Colorado, this time at the American research station of Point Barrow (71°N, 156°W) in Northern Alaska. Oltmans had noted, in particular the large day-to-day changes in O_3 during springtime.

This depletion of ozone to nearly zero levels was a surprise and its cause remained a mystery for several years. In 1988, however, a member of the Canadian research team, Len Barrie, noted that the ozone destruction was coincident with a rapid increase in the concentration of bromine species near the surface. He concluded that the springtime ozone destruction was due to the presence of high concentrations of halogen compounds, and was therefore probably caused by catalytic reactions involving BrO radicals.

What remained unclear was the origin of such large amounts of brominated species in this remote part of the world. The response to this unresolved question was found in 1992 by Fan and Daniel Jacob at Harvard University and by Jack Mc Connell at York University. These scientists showed that the observed "bromine explosion" was triggered by a sequence of reactions that removed bromide from surface sea ice and sea-salt aerosols, and released active bromine in the form of Br_2. As soon as solar radiation becomes available at polar latitudes in early spring, Br_2 is photolyzed and reactive Br and BrO are formed and rapidly destroy ozone.

CHAPTER ELEVEN

A Success Story

G radual and continuous scientific progress made since 1839 has allowed us to understand the nature and the role of ozone in the atmosphere, and specifically the ability of this gas to protect living beings from the detrimental action of solar ultraviolet radiation. From this historical presentation, we can retain some lessons on the importance of fundamental scientific research. Perhaps the first of them is that the production of new knowledge is often unforeseen: the discovery of ozone, arisen in a laboratory experiment on water electrolysis, was not planned under a preestablished project. Second, this unexpected discovery was followed by three decades of discussion among the best chemists of the time, and even generated heated debates and controversies among these scientists. As in many other disciplines—for example, the confrontation between Niels Bohr and Albert Einstein on their vision of quantum physics—research is often based on ideas that are first formulated as hypotheses but which, in order to be adopted, must be confirmed by experiments; these ideas must be challenged in the spirit of "free examination." Third, research on frontier topics that have potential consequences for the economic system can be disturbed by political mechanisms that have little to do with the scientific activity. Harold Johnston at the University of California, Berkeley describes how this situation may emerge:

> Scientist A publishes an article. Interest group B, with or without distorting this article, uses it to advance its cause and makes demands that conflict with the interests of group C. Group C hastily attacks A's person and motives. Both C and A feel outraged. Typically, neither B nor C understands the science of the original article.

Such conflicts have occurred on several occasions around questions related to the impact on ozone of human activities: potential operations of supersonic aircraft and space shuttle, use of chlorofluorocarbons, and rapid growth in automobile traffic. Scientific developments that have impinged on society have been the subject of lively debates and occasionally vigorous disagreements between scientists, politicians, and business representatives. Only a rigorous application of the scientific method has provided convincing arguments and led to some resolution of these conflicts.

Fourth, it will be recalled that the culture-sustaining research activities has evolved considerably during the historical period covered in this volume. The first discoveries were made by a few motivated and enthusiastic researchers who worked in isolation, sometimes used part of their private financial assets to sustain their research, and corresponded by letters with their foreign colleagues. The importance given to publications in peer-reviewed journals and the primary role played by the Academies are also noteworthy.

After the 1950s, the research's infrastructure evolved considerably as more advanced instrumentation became available by the rapid development of technology. The emergence of rockets and then satellites provided the opportunity to probe the atmosphere in unexplored regions and measure previously inaccessible quantities. Progress also resulted from the emergence of mathematical models whose complexity gradually increased and which required access to the most powerful supercomputers. The introduction of the modeling approach as a "third method" of scientific investigation (in between theory and experimentation) has had a considerable influence on atmospheric research since the 1960s. Models provide knowledge by establishing whether assumptions that are made by theories are verified by observations. In fact, fundamental progress has often resulted from the inability of models to reproduce reality, which led Nicolas Rouche (1925–2008) to claim that "the history of science is the history of inadequate models."

Finally, since the 1970s, the ozone problem has shifted from the academic world to that of policy-makers. The role of "curiosity-driven" research has often given way to the need to take swift and effective measures to preserve the very fragile global environment. Thanks to advances in research, decision-makers quickly decided that knowledge was sufficient to take the necessary measures to preserve the ozone layer. The Montreal Protocol on Substances that Deplete the Ozone Layer was signed in September 1987 and should be viewed as a landmark agreement for the protection of the planet and a success for the scientific community involved in ozone research (Figure 11.1). Despite this success, one should never forget, however, that no one is ever safe from real surprises, especially when it comes to a system as complex and as chaotic as the Earth's system: the appearance of a large hole in the ozone layer over Antarctica was not expected by anyone and took everybody by surprise. It generated new research that broadened the scope of our knowledge. The history of ozone research also showed the need for the scientific world to work in symbiosis with industry, policy-makers, and representatives of international organizations. The development of environmental diplomacy, which led to the ratification of a protocol on the protection of the ozone layer by almost all the world's nations, is a real success that, it is hoped, will be repeated on other environmental

Figure 11.1. A group of ozone scientists gathering in Athens, Greece, in September 2007 at the occasion of the 20th anniversary of the signature of the Montreal Protocol. Among the participants invited by the President of the International Ozone Commission, Christos Zerefos (in the middle of the front row), are Nobel Laureates Mario Molina and Sherry Rowland (first row, center) as well as Joe Farman who discovered the ozone hole (behind at the extreme right) and Shigeru Chubachi (next to Zerefos) who observed extreme low ozone amounts at the Japanese Antarctic station of Syowa. Courtesy: Christos Zerefos, Observatory of Athens, Greece.

issues affecting the future of our planet. We know today from the model simulations conducted at NASA by Paul Newman what would have happened in the future if the chlorofluorocarbons had not been regulated. Had the world allowed the emissions of chlorofluorocarbons to increase by 3% per year starting in 1974, about two thirds of the ozone would be destroyed by 2065.

After the ozone question, the problem of climate change is now being raised, and the systematic analysis of observations made for more than a century together with the projections made by the most advanced mathematical models lead us to a clear diagnosis: human activity contributes significantly to climate change. We know that the Earth will further warm up, polar ice melt, sea levels rise, and extreme weather events become more frequent. The majority of the states are now convinced by this message, but are we safe from the occurrence of geophysical events that would take by surprise our political leaders and the population of the Earth? More than ever, and despite the urgency of the problems to be solved, a dynamic effort in basic research must be maintained in order to preserve the intellectual capital that will help society address new challenges and respond to surprises that will inevitably occur in the future.

Selected Bibliography

Below is a sampling of key resource material used in the preparation of this book. It includes books, review papers, websites, selected individual studies, and key articles in the media relevant to each chapter.

References of General Interest

Andrews, D. G., C. B. Leovy, and J. R. Holton. *Atmosphere Dynamics* (San Diego, CA: Academic Press, 1987).

Bojkov, R. D. *The International Ozone Commission (IO3C). Its History and Activities Related to Atmospheric Ozone.* (Athens: Academy of Athens, Research Centre for Atmospheric Physics and Climatology, Publication 18, 2010).

Brasseur G., and S. Solomon. *Aeronomy of the Middle Atmosphere* (Dordrecht, The Netherlands: Springer, 2005), 644 pp.

Calvert, J. G., J. J. Orlando, W. R. Stockwell, and T. J. Wallington. *The Mechanisms of Reactions Influencing Atmospheric Ozone* (Oxford, UK: Oxford University Press, 2015), 608 pp.

Christie, M. *The Ozone Layer: A Philosophy of Science Perspective* (Cambridge, UK: Cambridge University Press, 2001), 215 pp.

Conway, E. M. *Atmospheric Science at NASA* (Baltimore, MD: The John Hopkins University Press, 2008), 386 pp.

Craig, R. A. "The Observations and Photochemistry of Atmospheric Ozone and their Meteorological Significance." *Meteorological Monographs* 1 (1959): 1–50.

Dessler, A. *Chemistry and Physics of Stratospheric Ozone* (New York: Academic Press, 2000), 214 pp.

Dobson, G. M. B. *Exploring the Atmosphere*, 2nd ed. (Oxford, UK: Oxford University Press, 1968), 209 pp.

Dotto, L., and H. Schiff. *The Ozone War*. (Doubleday & Co., 1978), 342 pp.

Engler, C. *Historisch-kritische Studien über das Ozon*, Leopoldina, Amtliches Organ der Kaiserlichen Leopoldinisch-Carolinischen Deutschen Akademie der Naturforscher, Heft XV, 1897, 36 pp.

Fabry, C. *L'ozone atmosphérique*, Éditions du Centre National de la Recherche Scientifique, 1950, 278 pp.

Finlayson-Pitts, B. J., and J. N. Pitts Jr. *Chemistry of the Upper and Lower Atmosphere* (Cambridge, MA: Academic Press, 2000), 969 pp.

Fleming, J. "Planetary-Scale Fieldwork: Harry Wexler on the Possibilities of Ozone Depletion and Climate Control." In *Knowing Global Environments: New Historical Perspectives on the Field Sciences*, ed. J. Vetter (New Brunswick: Rutgers University Press, 2011a), 190–211, 272 pp.

Fleming, J. R. *Fixing the Sky: The Checkered History of Weather and Climate Control* (New York: Columbia University Press, 2011b).

Fonrobert, E. *Das Ozon* (Stuttgart: Enke Verlag, 1916), 282 pp.

Fox, C. B. *Ozone and Antozone, Their History and Nature* (London: Churchill, 1873), 329 pp.

Goody, R. M. *The Physics of the Stratosphere*. (New York: Cambridge University Press, 1954), 187 pp.

Jacobson, M. Z. *Air Pollution and Global Warming: History, Science, and Solutions* (Cambridge, UK: Cambridge University Press, 2012), 370 pp.

Junge, C. E., C. W. Chagnon, and J. E. Manson. "Stratospheric aerosols." *Journal of Meteorology* 18 (1961): 81–108.

Lambright, W. H. *The Case of Ozone depletion, NASA and the Environment*, Monograph in Aerospace History, NASA-SP-2005-4538, 2005.

Latarjet, R. "Influence des variations de l'ozone atmosphérique sur l'activité biologique du rayonnement solaire." *Revue d'Optique Théorique et Instrumentale* 14 (1935): 398–414.

Leighton, P. J. *Photochemistry of Air Pollution* (New York: Academic Press, 1961), 312 pp.

Moeller, M. *Das Ozon*, Druck and Verlag von Friedr (Vieweg & Sohn, Braunschweig, 1921), 155 pp.

Müller, R. A brief history of stratospheric ozone research. *Meteorologische Zeitschrift* 18 (2009): 3–24.

———, ed. *Stratospheric Ozone Depletion and Climate Change.* (Cambridge, UK: RSC Publishing, 2011), 346 pp.

Oreskes, N., and E. M. Conway. *Merchants of Doubt. How a Handful of Scientists Obscured the Truth on Issues from Tobacco Smoke to Global Warming.* (New York: Bloomsbury Press, 2010), 356 pp.

Parson, E. A. *Protecting the ozone layer, Science and Strategy* (New York: Oxford University Press, 2003), 376 pp.

Seinfeld, J. H., and S. N. Pandis. *Atmospheric Chemistry and Physics, From Air Pollution to Climate Change* (New York: Wiley, 2006), 1152 pp.

Singh, O. N., and P. Fabian. *Atmospheric Ozone: A Millennium Issue.* European Geosciences Union, Special Publication Series, Vol. 1 (Göttingen, Germany: European Geosciences Union, 2003).

Sonnemann G. *Ozon: Natürliche Schwankungen und anthropogene Einflüsse.* (Berlin: Akademie Verlag, 1992).

Turco, R. P. "The photochemistry of the stratosphere." In *The Photochemistry of Atmospheres: Earth, the Other Planets, and Comets,* ed. J. S. Levine (Orlando, FL: Academic Press, 1985), 77–128.

Warneck, P. *Chemistry of the Natural Atmosphere* (San Diego, CA: Academic Press, 2000), 927 pp.

Chapter 1

Andrews, T. "On the origin of ozone." *Philosophical Transactions of the Royal Society of London* 6 (1865): 1–13.

Andrews, T. "Ozone study." *Proceedings of the Royal Society of London* XCIV (1867).

Andrews, T., and P. G. Tait. "On the volumetric relations of ozone and the action of the electric discharge on oxygen and other gases." *Philosophical Transactions of the Royal Society* 150 (1860): 113–32, https://doi.org/10.1098/rstl.1860.0008.

de la Rive, A. "Neue Untersuchungen über die Eigenschaften der discontinuirlichen elektrischen Ströme von abwechselnd entgegenesetzter Richtung, Parts 1 and 2." *Annual Review of Physical Chemistry* 54 (1840): 254 and 378–410.

de la Rive, A. Quelques Observations sur le Mémoire de M. Schoenbein relatif à la production de l'ozone par voie Chimique, Arch. élec. (Supp. à la Bibliothèque Universelle de Genève), 4 (1844): 454–56.

de la Rive, A. "Sur l'ozone." *Comptes rendus de l'Académie des Sciences* 20 (1845): 1291.

Engler, C., and O. Nasse. "Ozon und Antozon." *Annalen der Chemie und Pharmacie* 154 (1870): 215–37.

Kahlbaum, G. W. A., and F. V. Darbishire, eds. *The Letters of Faraday and Schoenbein, 1836–1862* (London: Benno Schwabe, Bâle and Williams and Norgate, 1899).

Leeds, A. R. "The history of antozone and peroxide of hydrogen." *Annals of the New York Academy of Sciences* 1 (1879): 405–25.

Marignac, C., and M. de la Rive. "Sur la production et la nature de l'ozone." *Comptes rendus de l'Académie des Sciences* 20 (1845): 808.

Meissner, G. C. F. Neue Untersuchungen über den elektrisierten Sauerstoff, Göttingen, bei Diederich, 1869.

Nolte, P. *Christian Friedrich Schönbein. Ein Leben für die Chemie 1799–1868*, 1999: Arbeitskreis Stadtgeschichte der Volkshochschule Metzingen-Ermstal e.V., ISBN 3-9802924-6-0.

Rubin, M. B. "The history of ozone. The Schönbein period, 1839–1868." *Bulletin for the History of Chemistry* 26 (2001): 40–56.

———. "The history of ozone. II. 1869–1899." *Bulletin for the History of Chemistry* 27 (2002): 81–106.

———. "The history of ozone, III, C. D. Harries and the introduction of ozone into organic chemistry." *Helvetica Chimica Acta* 86 (2003): 930–36.

———. "The history of ozone. IV. The isolation of pure ozone and determination of its physical properties." *Bulletin for the History of Chemistry* 29 (2004): 99–106.

———. "The history of ozone. V. Formation of ozone from oxygen at high temperatures." *Bulletin for the History of Chemistry* 32 (2007): 45–56.

———. "The history of ozone. VI. Ozone on silica gel (dry ozone)." *Bulletin for the History of Chemistry* 33 (2008): 68–75.

———. "The history of ozone. VII. The mythical spawn of ozone: Antozone, oxozone and ozohydrogen." *Bulletin for the History of Chemistry* 34 (2009): 39–49.

Schönbein, C. F. "Recherches sur la nature de l'odeur qui se manifeste dans certaines actions chimiques." *Comptes rendus de l'Académie des Sciences* 10 (1840a): 706–10.

———. "On the odour accompanying electricity and on the probability of its dependency on the presence of a new substance." *Philosophical Magazine* 17 (1840b): 293–94.

———. "Beobachtungen über den bei der Electrolysation des Wassers und dem Ausströmen der gewohnlichen Electricität aus Spitzen sich entwikkelnd en Geruch." *Annalen der Physik und Chemie (Poggendorf's Annalen)* 50 (1840c): 616.

———. "Sur la nature de l'ozône." *Archives de l'Électricité* 5 (1845): 11–23.

———. "Einige Bemerkungen uber die Anwesenheit des Ozons in der atmosphärischen Luft und die Rolle welcher dieser bei langsamen Oxydationen spielen dürfte." *Annalen der Physik und Chemie (Poggendorf's Annalen)* 65 (1845): 161–72.

———. "Über die Natur und den Namen des Ozons." *Journal für praktische Chemie* 56 (1852): 343–53.

————. "On the Various Conditions of Oxygen." *Philosophical Magazine IV* 15 (1858): 24–27.

————. "Über den freien postiv-activen Sauerstoff oder das Antozone." *Verhandlungen der natuforschenden Gesellschaft in Basel III* 2 (1860): 155–65.

Soret, J. L. "Sur la production de l'ozone par l'électrolyse et sur la nature de ce corps." *Comptes rendus hebdomadaires des séances de l'Académie des sciences, Series C* 56 (1863): 390–93.

————. "Recherches sur la densité de l'ozone." *Comptes rendus hebdomadaires des séances de l'Académie des sciences, Series C* 61 (1865): 941.

Soret, J. L. "Recherches sur la densité de l'ozone." *Annales de Chimie et de Physique* 4/VII (1866): 113-118 and 4/XIII (1868): 257–282.

von Babo, L. "Wird neben ozon durch den electrischen Strom auch sogenanntes antozon erzeugt?" *Annalen der Chemie und Pharmacie. II Supp.* (1863): 291–96.

Weltzien, C. "Sur la polarisation de l'oxygène, les ozonides et les antozonides." *Annalen der Chemie und Pharmacie* 59 (1860): 105–110.

————. "Über das Wasserstoffperoxyd und das Ozon." *Annalen der Chemie und Pharmacie* 138 (1866): 129–64.

Chapter 2

Albert Lévy. *Histoire de l'Air*, Librairie Germer Baillière et Cie, Paris, 1879.

————. "Analyse de l'air atmosphérique—Ozone." *Annales de l'Observatoire Municipal de Montsouris* 8 (1907): 289–91.

Harries, C. D. "Über die Einwirkung des Ozons auf organische Verbindungen. (Erste Abhandlung)." *Justus Liebigs Annalen der Chemie* 343 (1905): 311–75.

Houzeau, A. "Recherches sur l'oxygène à l'état naissant." *Comptes rendus hebdomadaires des séances de l'Académie des sciences, Series C* 40 (1850): 947–50 and C 43 (1856): 34–38.

————. "Méthode analytique pour reconnaître et doser l'oxygène naissant." *Comptes rendus hebdomadaires des séances de l'Académie des sciences, Series C* 45 (1857): 873–77.

————. "Preuve de la présence dans l'atmosphère d'un nouveau principe gazeux-l'oxygen naissant." *Comptes rendus de l'Académie des Sciences* 46 (1858): 89.

————. "Sur l'ozone atmosphérique," *Annales de Chimie et de Physique* 27 (1872): 5–68.

Linvill, D. E., W. J. Hooker, and B. Olson. "Ozone in Michigan's environment 1876–1880." *Monthly Weather Review* 108 (1980): 1883–1891.

Schöne, J. L. "Über atmosphärisches Ozon." *Nature* 29 (1863): 617–18.

Tarasick, D., I. E. Gabally, O. R. Cooper, M. G. Schultz, G. Ancellet, T. Leblanc, T. J. Wallington, J. Ziemke, X. Liu, M. Steinbacher, J., Staehelin, C. Vigouroux, J. W. Hannigan, O. García, G. Foret, P. Zanis, E. Weatherhead, I. Petropavlovskikh, H. Worden, M. Osman, J. Liu, K.-L. Chang, A. Gaudel, M. Lin, M. Granados-Muñoz, A. M. Thompson, S. J. Oltmans, J. Cuesta, G. Dufour, V. Thouret, B. Hassler, T. Trickl, and J. L. Neu, "Tropospheric ozone assessment report: Tropospheric ozone from 1877 to 2016, observed levels, trends and uncertainties." *Elementa: Science of the Anthropocene* 7(1) (2019): 39. DOI: http://doi.org/10.1525/elementa.376.

Volz A., and D. Kley. "Evaluation of the Montsouris Series of Ozone Measurements made at the Nineteenth Century." *Nature* 332 (1988): 240–42.

Chapter 3

Brönnimann, S., J. Staehelin, S. F. G. Farmer, J. C. Caine, T. Svendby, and T. Svenøe. "Total ozone observations prior to the IGY. I: A history." *Quarterly Journal of the Royal Meteorological Society* 129 (2003): 2797–817.

Chappuis, J. "Sur le spectre d'absorption de l'ozone." *Comptes rendus de l'Académie des Sciences* 91 (1880): 985–86.

Cornu, A. "Sur la limite ultra-violette du spectre solaire." *Comptes rendus de l'Académie des Sciences* 88 (1879): 1101–08.

Crutzen, P. "Ozone production rates in an oxygen-hydrogen-nitrogen oxide atmosphere." *Journal of Geophysical Research* 76 (1971): 7311–27.

——. "A review of upper atmospheric photochemistry." *Canadian Journal of Chemistry* 52 (1974): 1569–81.

Dobson, G. M. B. "Measurements of the Sun's ultra-violet radiation and its absorption in the Earth's atmosphere." *Proceedings of the Royal Society of London* 104A (1923): 252–71.

Dobson, G. M. B. "Atmospheric ozone." *Gerlands Beiträge zur Geophysik* 24 (1929): 8–15.

Dobson, G. M. B. "A photoelectric spectrophotometer for measuring the amount of atmospheric ozone." *Proceedings of the Physical Society* 43 (1931): 324–39.

Dobson, G. M. B. "Forty years of ozone research at Oxford: A history." *Applied Optics* 7 (1968): 387–405.

Dobson, G. M. B., and D. N. Harrison. "Measurements of the amount of ozone in the Earth's atmosphere and its relation to other geophysical conditions." *Proceedings of the Royal Society of London, A* 110 (1926): 660–93.

Dorno, C. "Über Ozonmessungen auf spektroskopischem Wege" *Meteorologische Zeitschrift* 44 (1927): 385–390 (including Reply to Götz).

Dütsch, H. U. "The ozone distribution in the atmosphere." *Canadian Journal of Chemistry* 52 (1974): 1491–504.

Fabry, C., and H. Buisson. "L'absorption de l'ultra-violet par l'ozone et la limite du spectre solaire." *Journal of Theoretical and Applied Physics* 3 (1913): 196–206.

———. "Étude de l'extrémité ultra-violette du spectre solaire." *Journal de Physique et Le Radium* 2 (1921): 197–226.

Féry, C. "The curved prism spectrograph." *Applied Journal of Paris* 34 (1911): 79.

Fowler, A., and J. Strutt. "Absorption bands of atmospheric ozone in the spectra of the sun and stars." *Proceedings of the Royal Society of London, A* 93 (1917): 577–86.

Götz, F. W. P. "Der Jahresgang des Ozongehaltes der hohen Atmosphäre." *Beitrage Physics Atmosphere* 13 (1926): 15–22.

———. "Das Atmospherische Ozon." *Ergebnisse der Kosmischen Physik* 1 (1931): 180–235.

———. "Ozone in the atmosphere." In Malone, T. F. (ed.). *Compendium of Meteorology*, (Boston: American Meteorological Society, 1951) 275–291.

Götz, F. W. P., and G. M. B. Dobson. "Observations of the height of the ozone in the upper atmosphere, Vertical distribution of ozone in the atmosphere." *Proceedings of the Royal Society of London* A 120 (1928 and 1929): 251–59; and A 125: 292–94.

Götz, F. W. P., G. M. B. Dobson, and A. R. Meetham. "Vertical distribution of ozone in the atmosphere." *Nature* 132 (1933): 281.

Götz, F. W. P., A. R. Meetham, and G. M. B. Dobson. "The vertical distribution of ozone in the atmosphere." *Proceedings of the Royal Society of London* 145A (1934): 416–46.

Gushtin, G. P. "Universal ozonometer." *Proceeding of the Main Geophysical Observatory* 141 (1963): 83–8; Leningrad.

Gushtin, G. P., and S. A. Sokolenko, "The improved 3-wavelenghts (300, 326, and 348 ± 2 nm) ozonometer." *Proceeding of the Main Geophysical Observatory* 472 (1984): 35–40; Leningrad.

Hartley, W. N. "On the absorption of solar rays by atmospheric ozone." *Journal of the Chemical Society, Transactions* 39 (1881a): 111–28.

———. "On the absorption spectrum of ozone." *Journal of the Chemical Society* 39 (1881b): 57.

Huggins, W., and Mrs Huggins. "On a new group of lines in the photographic spectrum of Sirius." *Proceedings of the Royal Society of London* 48 (1890): 216–17.

Lejay, P. "Ozone measurements and currents in the stratosphere." *Bulletin of the American Meteorological Society* 20 (1939): 298–301.

Meetham, A. R. "The correlation of the amount of ozone with other characteristics of the atmosphere." *Quarterly Journal of the Royal Meteorological Society* 63 (1937): 289–307.

Perl, G. "Das bodennahe Ozon in Arosa: seine regelmässigen und unregelmässigen Schwankungen." *Archiv für Meteorologie, Geophysik und Bioklimatologie* 14 (1965): 449–58.

Soret, J.-L. "Note sur les relations volumétriques de l'ozone." *Comptes rendus de l'Académie des Sciences.* 62 (1863a): 608.

———. "Sur les relations volumétriques de l'ozone." *Comptes rendus de l'Académie des Sciences* 57 (1863b): 604–09.

———. "Recherches sur la densité de l'ozone." *Comptes rendus de l'Académie des Sciences* 61 (1865): 941.

Staehelin, J., S. Brönnimann, T. Peter, R. Stübi, P. Viatte, and F. Tummon. "The value of Swiss long-term ozone observations for international atmospheric research." In *From Weather Observations to Atmospheric and Climate Sciences in Switzerland—Celebrating 100 years of the Swiss Society for Meteorology*, eds. S. Willemse and M. Fürger (vdf Hochschulverlag AG and der ETH Zürich, 2016), 325–49.

Staehelin, J., J. Thudium, R. Buehler, A. Volz-Thomas, and W. Graber. "Trends in surface ozone concentrations at Arosa (Switzerland)." *Atmospheric Environment* 28 (1994): 75–88.

Staehelin, J., P. Viatte, R. Stübi, F. Tummon, and T. Peter. "Stratospheric ozone measurements at Arosa (Switzerland): History and scientific relevance." *Atmospheric Chemistry and Physics* 18 (2018): 6567–6584.

Strutt, R. J. "Ultra-violet transparency of the lower atmosphere, and its relative poverty in ozone." *Proceedings of the Royal Society of London* 94A (1918): 260–68.

Trenkel, H. Prof. Dr. Paul Götz, Falkenstein: *Zeitschrift der Studentenverbindungen Schwizerhusli Basel*, Zähringia Bern, Carolingia Zurich, Valdesia, Lausanne, 1954, 199–201.

Vassy, A. and E. Vassy. Etude de l'ozone dans ses rapports avec la circulation atmosphérique. *La Météorologie* 19 (1939): 3–16.

Walshaw, C. D. "G.M.B. Dobson—The man and his work." *Planetary Space Science* 37 (1989): 1485–507.

Chapter 4

Chapman, S. "A theory of upper atmospheric ozone." *Memoirs of the Royal Meteorological Society* 3 (1930a): 103–25.

———. On ozone and atomic oxygen in the upper atmosphere. *London Edinburgh and Dublin Philosophical Magazine and. Journal of. Science 7* 10 (1930b): 369–83.

Dütsch, H. U. "The photochemistry of stratospheric ozone." *Quarterly Journal of the Royal Meteorological Society* 94 (1968): 483–97.

Nicolet, M., and W. Peetermans. "Atmospheric absorption in the O2 Schumann-Runge band spectral range and photodissociation rates in the stratosphere and mesosphere." *Planetary and Space Science* 28 (1980): 84–103.

Schumacher, H. J. "The mechanism of photochemical decomposition of ozone" *Journal of the American Chemical Society* 52 (1930): 2377–391.

Chapter 5

Anderson G. P. et al. "Satellite observations of the vertical ozone distribution in the upper stratosphere." *Annales de Géophysique* 25 (1969): 341–45.

Assmann, R. "Über die Existenz eines wärmeren Luftstromes in der Höhe von 10 bis 15 km." *Preußische Akademie der Wissenschaften, Physikalisch-Mathematische Klasse* 24 (1902): 1–10.

Cabannes, J., and J. Dufay. "Mesure de l'altitude de la couche d'ozone dans l'atmosphère." *Comptes rendus de l'Académie des Sciences* 181 (1925): 302–04.

Chalonge, D. "Sur la repartition de l'ozone dans l'atmosphère terrestre." *Journal of Physics* 1 (1932): 21–42.

Chalonge, D., and F. W. P. Götz. 1929. "Mesures diurnes et nocturnes de la quantité d'ozone contenue dans la haute atmosphère." *Comptes rendus de l'Académie des Sciences* 189 (1929): 704–06.

Coblentz, W. W., and R. Stair. "Distribution of ozone in the stratosphere: Measurements of 1939 and 1940." *Journal of Research of the National Bureau of Standards* 26 (1941): 161–74.

Dave J. V., and C. L. Mateer. "A preliminary study of the possibility of estimating total atmospheric ozone from satellite measurements." *Journal of the Atmospheric Sciences* 24 (1967): 414–27.

De Bort, L. P. T. "Variations de la température de l'air libre, dans la zone comprise entre 8 et 15 kilomètres d'altitude." *Comptes rendus de l'Académie des Sciences* 134 (1902): 987–89.

Burrows, J. P. et al. "The Global Ozone Monitoring Experiment (GOME): Mission concept and first scientific results." *Journal of the Atmospheric Sciences.* 56 (1999): 151–75.

Dütsch, H. U. "The ozone distribution in the atmosphere." *Canadian Journal of Chemistry* 52 (1974): 1491–504.

Iozenas, V. A. "Determination of the vertical ozone distribution in the upper layers of the atmosphere from satellite measurements of ultraviolet solar radiation Scattered by the Earth's atmosphere." *Geomagnetism and Aeronomy* 8 (1968): 403–07.

Iozenas V. A. et al. "An investigation of the planetary ozone distribution from satellite measurements of ultraviolet spectra." *Izvestia, Atmospheric and Oceanic Physics* 5 (1969): 219–33.

Johnson, F. S., J. D. Purcell, and R. Tousey. "Measurements of the vertical distribution of atmospheric ozone from rockets." *Journal of Geophysical Research* 56 (1951): 583–94.

Johnson, F. S., J. D. Purcell, R. Tousey, and K. Watanabe. "Direct measurements of the vertical distribution of atmospheric ozone to 70 kilometers altitude." *Journal of Geophysical Research* 57 (1952): 157–76.

Krueger, A. J. "The mean ozone distributions from several series of rocket soundings to 52 km at latitudes from 58°S to 64°N." *Pure and Applied Geophysics* 106–08 (1973): 1271–80.

Mani, A. "Ozone studies in India." *Indian Journal of Radio & Space Physics* 19 (1990): 542–49.

Mateer, C. L., D. F. Heath, and A. J. Krueger. "Estimation of total ozone from satellite measurements of backscattered ultraviolet earth radiance." *Journal of the Atmospheric Sciences* 28 (1971): 1307–11.

Miller, A. J. "A review of satellite, observations of atmospheric ozone." *Planetary and Space Sciences* 37 (1989): 1539–54.

Piccard, A. *Au-dessus des nuages* (Paris: Editions Grasset, 1933).

Ramanathan, K. R., and R. N. Kulkarni. "Mean meridional distributions of ozone in different seasons calculated from *umkehr* observations and probable vertical transport mechanisms." *Quarterly Journal of the Royal Meteorological Society* 86 (1960): 144–55.

Rawcliffe, R. D., and D. D. Elliott. "Latitude distributions of ozone at high altitudes deduced from satellite measurement of the earth's radiance at 2840 Å." *Journal of Geophysical Research* 71 (1966): 5077–89.

Regener, E., and V. H. Regener. "Aufnahme des ultravioletten Sonnenspektrums in der Stratosphäre und vertikale Ozonverteilung." *Zeitschrift für Physik.* 35 (1934): 788–93.

Remsberg, E. E., J. M. Russell III, J. C. Gille, L. L. Gordley, P. L. Bailey, W. G. Planet, and J. E. Harries. "The validation of NIMBUS 7 LIMS measurements of ozone." *Journal of Geophysical Research* 89 (1984): 5161–78.

Singer, S. F., and R. C. Wentworth. "A method for the determination of the vertical ozone distribution from a satellite." *Journal of Geophysical Research* 62 (1957): 299–308.

Chapter 6

Bates D. R., and M. Nicolet. "The photochemistry of atmospheric water vapor." *Journal of Geophysical Research* 55 (1950): 301–26.

Brewer, A. W. "Evidence for a world circulation provided by measurements of helium and water vapour distribution in the stratosphere." *Quarterly Journal of the Royal Meteorological Society* 75 (1949): 351–63.

Brix, P. A., and G. Herzberg. "The dissociation energy of oxygen." *The Journal of Chemical Physics* 21 (1953): 2240–41.

Crutzen, P. "The influence of nitrogen oxides on the atmospheric ozone content." *Quarterly Journal of the Royal Meteorological Society* 96 (1970): 320–25.

———. "Ozone production rates in an oxygen-hydrogen-nitrogen oxide atmosphere." *Journal of Geophysical Research* 76 (1971): 7311–27.

———. "A review of upper atmospheric photochemistry." *Canadian Journal of Chemistry* 52 (1974): 1569–81.

Cunnold, D., F. Alyea, N. Phillips, and R. Prinn. "A three-dimensional dynamical-chemical model of atmospheric ozone." *Journal of the Atmospheric Sciences* 32 (1975): 170–94.

Garcia, R. R., and S. Solomon. "A new numerical model of the middle atmosphere: 2. Ozone and related species." *Journal of Geophysical Research* 99 (1992): 12937–51.

Garcia, R. R., F. Stordal, S. Solomon, and J. T. Kiehl. "A new numerical model of the middle atmosphere: 1. Dynamics and transport of tropospheric source gases." *Journal of Geophysical Research* 97 (1992): 12967–91.

Hampson, J. *Photochemical behaviour of the ozone layer.* Tech. Note 1627, Canadian Arm. Research and Develop. Establishment, Quebec, 1964.

Holton, J. R. "On the global exchange of mass between the stratosphere and troposphere." *Journal of the Atmospheric Sciences* 47 (1990): 392–95.

Holton, J. R., P. H. Haynes, M. E. McIntyre, A. R. Douglas, R. B. Rood, and L. Pfister. "Stratosphere troposphere exchange." *Review of Geophysics* 33 (1995): 403–39.

Hunt, B. G. "Photochemistry of ozone in a moist atmosphere." *Journal of Geophysical Research* 71 (1966): 1385–98.

———. "Experiments with a stratospheric general circulation model III. Large-scale diffusion of ozone including photochemistry." *Monthly Weather Review* 97 (1969): 297–306.

McIntyre, M. E. "On global-scale atmospheric and oceanic circulations." In *Perspectives in Fluid Dynamics*, eds. G. K. Batchelor, H. K. Moffatt, and M. G. Worster (Cambridge, UK: Cambridge Univ. Press, 2000).

McIntyre, M. E., and T. N. Palmer. "Breaking planetary waves in the stratosphere." *Nature* 305 (1983): 593–600.

Mecke, R. "The photochemical ozone equilibrium in the atmosphere." *Transactions of the Faraday Society* 27 (1931): 375–77.

Meinel, A. B. "Identification of the 6560 Å emission in the spectrum of the night sky." *The Astrophysical Journal* 111 (1950): 433–34.

Murcray, D. G., T. G. Kyle, F. H. Murcray, and W. J. Williams. "Nitric acid and nitric oxide in the lower stratosphere." *Nature* 218 (1968): 78–79.

Nicolet, M. "Stratospheric ozone: An introduction to its study." *Review of Geophysics* 13 (1975): 593–636.

Wulf, O. R. "The distribution of atmospheric ozone." *Bull. Meteorol. Soc.* 23 (1940): 439–46.

Wulf, O. R., and L. S. Deming. "The theoretical calculations of the distribution of photochemically-formed ozone in the atmosphere." *Journal of Geophysical Research* 41 (1936): 299–310.

———. "The distribution of atmospheric ozone in equilibrium with solar radiation and the rate of maintenance of the distribution." *Journal of Geophysical Research* 42 (1937): 195–202.

Chapter 7

Foley, H. M., and M. A. Ruderman. "Stratospheric NO production from past nuclear explosions." *Journal of Geophysical Research* 78 (1973): 4441–50.

Johnston, H. "Reduction of stratospheric ozone by nitrogen oxide catalysts from supersonic transport exhaust." *Science* 173 (1971): 517–22.

Johnston, H. S. "Atmospheric ozone." *Annual. Review of Physical Chemistry* 43 (1992): 1–32.

London, J., and J. Park "Application of general circulation models to the study of stratospheric ozone." *Pure Appl. Geophys.* 106–108 (1973): 1611–17.

Chapter 8

Benedick, R. E. *Ozone Diplomacy* (Cambridge, MA: Harvard University Press, 1998), 480 pp.

Fahey, D. "The Montreal protocol protection of ozone and climate." *Theoretical Inquiries in Law* 14 (2013): 21–42.

Fahey, D. W., P. A. Newman, J. A. Pyle, and B. Safari, eds. Scientific assessment of ozone depletion: 2018. Global Ozone Research and Monitoring Project Report 58, WMO, 2019, 590 pp.

Lovelock, J. E. "Atmospheric halocarbons and stratospheric ozone." *Nature* 252 (1974): 292.

Molina, M. J., and F. S. Rowland. "Stratospheric sink for chlorofluoromethanes: Chlorine atom-catalysed destruction of ozone." *Nature* 249 (1974): 810–12.

Prather, M. J., M. B. McElroy, and S. C. Wofsy. "Reductions in ozone at high concentrations of stratospheric halogens." *Nature* 312 (1984): 227–31.

Staehelin, J., N. R. P. Harris, C. Appenzeller, and J. Eberhard. Ozone trends: A review. *Review of Geophysics* 39 (2011): 231–90.

Stolarski, R. S., and R. J. Cicerone. "Stratospheric chlorine: A possible sink for ozone." *Canadian Journal of Chemistry* 52 (1974): 1610–15.

Chapter 9

Chapman, S. "Clouds high in the stratosphere." *Nature* 129 (1932): 497–99.

Chubachi, S. "A special ozone observation at Syowa station, Antarctica from February 1982 to January 1983." In *Atmospheric ozone*, eds. C. S. Zerefos and A. Ghazi (Norwell, MA: D. Reidel, 1985), 285–89.

Crutzen, P J., and Arnold, F. "Nitric acid cloud formation in the cold Antarctic stratosphere: A major cause for the springtime 'ozone hole." *Nature* 342 (1986): 651–55.

Douglass, A. R., P. A. Newman, and S. Solomon. "The Antarctic ozone hole: An update." *Physics Today* 67 (2014): 42–48.

Farman, J. C., B. G. Gardiner, and J. D. Shanklin. "Large losses of total ozone in Antarctica reveal seasonal ClOx/NOx interaction." *Nature* 315 (1985): 207–10.

Hofmann, D. J. "Balloon-borne measurements of middle atmosphere aerosols and trace gases in Antarctica." *Review of Geophysics* 26 (1988): 113–30.

Hofmann, D. J., and S. Solomon. "Ozone destruction through heterogeneous chemistry following the eruption of El Chichón." *Journal of Geophysical Research* 94 (1989): 5029–41.

McElroy, M. B., R. J. Salawitch, S. C. Wofsy, and J. A. Logan. "Reductions of Antarctic ozone due to synergistic interactions of chlorine and bromine." *Nature* 321 (1986): 759–62.

Molina, L. T., and M. J. Molina. "Production of Cl_2O_2 from the self-reaction of the ClO radical." *The Journal of Physical Chemistry* 91A (1987): 433–36.

Molina, M. J., T.-L. Tso, L. T. Molina, and F. C.-Y. Wang. "Antarctic stratospheric chemistry of chlorine nitrate, hydrogen chloride, and ice: Release of active chlorine." *Science* 238 (1987): 1253–57.

RealClimate.org. *Climate Science from Climate Scientists* (2017). http://www.realclimate.org/index.php/archives/2017/12/what-did-nasa-know-and-when-did-they-know-it/.

Solomon S. "The discovery of the Antarctic ozone hole." *Nature*, News and Views, 23 October 2019.

Solomon, S. "Stratospheric ozone depletion: A review of concepts and history." *Reviews of Geophysics* 37 (1999): 275–316.

Solomon, S., R. R. Garcia, F. S. Rowland, and D. J. Wuebbles. "On the depletion of Antarctic ozone." *Nature* 321 (1986): 755–58.

Taubes, G. "The ozone backlash." *Science* 260 (1993): 1580–83.

Toon, O. B., R. P. Turco, J. Jordan, J. Goodman, and G. Ferry. "Physical processes in polar stratospheric ice clouds." *Journal of Geophysical Research* 94 (1989): 359–80.

Chapter 10

Albert-Lévy. Ozone, Annuaire de l'Observatoire de Montsouris pour l'an 1877 (1877): 398–405.

———. "Analyse chimique de l'air—Ozone." *Annuaire de l'Observatoire de Montsouris* (1878): 495–505.

———. "Analyse de l'air atmosphérique—Ozone." *Annuaire de l'Observatoire de Montsouris* 8 (1907): 289–91.

Albert-Lévy, H. Henriet, and M. Bouyssy. L'ozone atmosphérique, *Annuaire de l'Observatoire Municipal de Montsouris* 6 (1905): 18–22, 315–21.

Barrie, L., and U. Platt. "Arctic tropospheric chemistry: an overview." *Tellus* 49B (1997): 450–54.

Blacet, F.E. "Photochemistry in the lower atmosphere." *Industrial & Engineering Chemistry* 44 (1952): 1339–42.

Bojkov, R. D. "Surface ozone during the second half of the nineteenth century." *Journal of Climate and Applied Meteorology* 25 (1986): 343–52.

Chameides, W., and D. H. Stedman. "Tropospheric ozone. Coupling transport and photochemistry." *Journal of Geophysical Research* 82 (1977): 1787–94.

Chameides, W., and J. C. G. Walker. "A photochemical theory of tropospheric ozone." *Journal of Geophysical Research* 78 (1973): 8751–60.

Crutzen, P. J. "Photochemical reactions initiated by and influencing ozone in unpolluted tropospheric air." *Tellus* 26 (1973): 47–57.

Fishman, J., and P. J. Crutzen. "The origin of ozone in the troposphere." *Nature* 274 (1978): 855–58.

Fishman, J., S. Solomon, and P. J. Crutzen. "Observational and theoretical evidence in support of a significant in-situ photochemical source of tropospheric ozone." *Tellus* 31 (1979): 432–46.

Götz, F. W. P., and F. Volz. "Aroser Messungen des Ozongehalts der unteren Troposphäre und sein Jahresgang." *Zeitschrift für Naturforschung* A6 (1951): 634–39.

Haagen-Smit, A. J. "Chemistry and physiology of Los Angeles smog." *Industrial & Engineering Chemistry* 44 (1952): 1342–46.

Haagen-Smit, A. J., C. E. Bradley, and M. M. Fox. "Ozone formation in photochemical oxidation of organic substances." *Industrial & Engineering Chemistry* 45 (1953): 2086–89.

Jacob, D., J. A. Logan, G. M. Gardner, R. M. Yevich, C. M. Spivakovsky, S. C. Wofsy, S. Sillman, and M. J. Prather. "Factors regulating ozone over the United States and its export to the global atmosphere." *Journal of Geophysical Research* 98 (1993): 14817–26.

Kley, D., A. Volz, and F. Mulheims. "Ozone measurements in historic perspective," in *Tropospheric Ozone*, ed. I. S. A. Isaksen, NATO ASI Series, Vol. 227 (Dordrecht, The Netherlands: D. Reidel, 1988), 63–72.

Levy, H. "Normal atmosphere: Large radical and formaldehyde concentrations predicted." *Science* 173 (1971): 141–43.

Logan, J. A., M. J. Prather, S. C. Wofsy, and M. B. McElroy. "Tropospheric chemistry: A global perspective." *Journal of Geophysical Research* 86 (1981): 7210–54.

Monks, P. J. et al. "Tropospheric ozone and its precursors from the urban to the global scale from air quality to short-lived climate forcer." *Atmospheric Chemistry and Physics* 15 (2015): 8889–973.

Oltmans, S. J. et al. "Recent tropospheric ozone changes—A pattern dominated by slow or no growth." *Atmospheric Environment* 67 (2013): 331–51.

Ridley, B. A., M. A. Carroll, G. L. Gregory, and G. W. Sachse. "NO and NO2 in the troposphere: Technique and measurements in regions of a folded troposphere." *Journal of Geophysical Research* 93 (1988): 15813–30.

Royal Society. *Ground-level ozone in the 21st century: Future trends, impacts and policy implications.* RS Policy Doc. 15/08 2008, 148 pp.

Thompson A. M., J. C. Witte, R. D. Hudson, Hua Guo, J. R. Herman, and M. Fujiwara. "Tropical tropospheric ozone and biomass burning." *Science* 291 (2001): 2128–32. doi:10.1126/science.291.5511.2128.

Thompson, A. M., J. C. Witte, R. D. McPeters, S. J. Oltmans, F. J. Schmidlin, J. A. Logan, M. Fujiwara, V. W. J. H. Kirchhoff, F. Posny, G. J. R. Coetzee, B. Hoegger, S. Kawakami, T. Ogawa, B. J. Johnson, H. Vömel, and G. Labow. "Southern hemisphere additional ozonesondes (SHADOZ) 1998–2000 tropical ozone climatology 1. Comparison with Total Ozone Mapping Spectrometer (TOMS) and ground-based measurements." *Journal of Geophysical Research* 108 (2003): PEM10 1–19.

Winkler Dawson, K. *Death in the Air* (Hachette, 2017).

Chapter 11

Godin-Beekmann, S., P. A. Newman, and I. Petropavlovskikh. "30th anniversary of the Montreal Protocol: From the safeguard of the ozone layer to the protection of the Earth's climate." *Comptes Rendus Geoscience* 350 (2018): 331–33.

Newman, P. A., L. D. Oman, A. R. Douglass, E. L. Fleming, S. M. Frith, M. M. Hurwitz, S. R. Kawa, C. H. Jackman, N. A. Krotkov, E. R. Nash, J. E. Nielsen, S. Pawson, R. S. Stolarski, and G. J. M. Velders "What would have happened to the ozone layer if chlorofluorocarbons (CFCs) had not been regulated?" *Atmospheric Chemistry and Physics* 9 (2009): 2113–28.

Zerefos C., G. Contopoulos, and G. Skalkeas, Gregory (Eds.). "Ozone, Twenty years of ozone decline." *Proceedings of the Symposium on 20th Anniversary of the Montreal Protocol.* (Dordrecht, The Netherlands: Springer, 2009), 470 pp.

Index

chlorofluorocarbons (CFCs), 5, 185,
208, 209
CFC-11, 171, 171n2, 173
CFC-12, 173
CFC Ozone Puzzle, 189
degradation of, 175
effects of, 169–181, 205
emissions, 210
emissions of, 232
hypothesis, 200
in ozone depletion, 198
in stratosphere, 209
cholera, 40
Chubachi, Shigeru, 1, 184*f*, 231*f*
Churchill, Winston, 215
CIAP. *See* Climatic Impact Assessment
Program (CIAP)
Cicerone, Ralph, 136, 137, 137*f*, 151, 167, 169,
202, 203
circulation models, 151
Clarke, Sir James, 30*f*
Clark, M. A., 114
Clausius, Rudolph, 17
Climatic Impact Assessment Program
(CIAP), 160, 162–165*b*
climatology
of stratospheric ozone, 118
of tropospheric ozone
column, 224*f*
cloud of fog and pollution, 216*f*
Coblentz, W. W., 104*f*, 105
Cold War, 3*b*, 168
Colombel, Augustin, 67
COMESA. *See* Committee on
Meteorological Effects
of Stratospheric Aircraft
(COMESA)
Comité pour l'étude des conséquences
des vols stratosphériques
(COVOS), 162
Comité Spécial de l'Année Géophysique
Internationale (CSAGI), 2*b*, 2*f*
Committee on Meteorological Effects
of Stratospheric Aircraft
(COMESA), 162

Commonwealth Scientific and Industrial
Research Organisation
(CSIRO), 38
Community Atmospheric Model with
Chemistry (CAM-Chem), 226
Concorde, 153, 154
Concorde aircraft 500, 162
condensation, 214
Conte, Silvio Ottavio, 154
contrarians, 205*b*
Conway, Erik, 207*b*
cooling agents, aerosols as, 155
Cornu, Marie Alfred, 47, 48
Coroniti, Samuel, 164*b*
cosmic radiation, 91
cosmic rays, 98
Cosyns, Max, 96
Coulomb, Jean, 2*f*
Crutzen, Paul, 130, 131, 131*f*, 132, 132n9, 133*f*,
135–137, 151, 159, 159n7, 176, 197,
198, 202, 222, 226
CSAGI. *See* Comité Spécial de l'Année
Géophysique Internationale
(CSAGI)
CSIRO. *See* Commonwealth Scientific and
Industrial Research Organisation
(CSIRO)
Cunnold, Derek, 148, 149*f*
cyclones, tropical, 131

Dahlem Conferences, 202n6
Danielson, Edwin F., 156, 157
data assimilation techniques, 227
Dave, J. V., 114, 114*f*, 115
Dawson, Kate Winkler, 216*f*
degradation, of chlorofluorocarbons, 175
de la Rive, Arthur, 11, 11*f*, 14, 16
de Marignac, Jean Charles Galissard,
11*f*, 14, 16
Deming, Lola S., 120, 120*f*, 121, 121*b*,
122*f*, 123
Deming, William Edwards, 121, 121*b*
Deniers, 205*b*–208*b*, 209
Denver group, 133, 134*f*

McNary, Robert R., 173
McNeal, Joseph (Joe), 225
Mecke, Reinhard, 120, 123
Meetham, A. R., 65, 86, 87, 88*b*, 89*f*
Mégie, Gérard, 180*f*
Meinel, Aden, 127
Meissner, Georg C. F., 16, 17*f*, 19, 20*b*, 21
meridional transport, in stratosphere, 141*f*
metaphor, 191*b*
Meteorological satellite (MetOp), 118
M-83 filter instrument, 69
Microwave Limb Sounder (MLS), 116
microwave spectrometer, 200
middle atmosphere, multi-dimensional
        chemical-transport modeling
        of, 149*f*
Midgley, Thomas J. Jr., 172, 173
Mitchell, A., 44
MLS. *See* Microwave Limb Sounder (MLS)
Moff at, W. F., 30*f*, 30n3
MOGUNTIA, 225–226
moist stratosphere, 128
molecular oxygen, and ozone,
        121–122, 122*f*
Molina, Luisa Tan, 203, 204*f*
Molina, Mario, 169, 170, 170*f*, 174, 175*f*, 203,
        204*f*, 231*f*
Montreal Protocol, 177*b*, 178–179, 181*f*, 193,
        209, 231, 231*f*
MOZART model, 226
Muller, Jean-Francois, 226
Murcray, David G., 132
Murgatroyd, Robert J., 162

NASA (National Aeronautics and Space
        Administration (NASA)), 1, 62,
        96, 137, 167, 183, 188
    Earth Observing System (EOS),
        117*f*, 118
    ER-2 aircraft, 201*f*
    Global Tropospheric Experiments
        Program, 225*f*
    Goddard Space Flight Center, 137, 189*b*
    GTE program, 226

Orbiting Geophysical Observatory
        (OGO, 1967–1969), 114
Ozone Processing Team, 188
    satellite images, 193
Upper Atmosphere Research Satellite
        (UARS), 116
nascent oxygen *(oxygène naissant)*, 12n5
Nasse, Otto Johann, 21
Natarajan, Murali, 196, 197
National Center for Atmospheric Research
        (NCAR), Boulder, 113–115, 156
National Oceanic and Atmospheric
        Administration (NOAA), 171
National Ozone Expedition (NOZE), 200
natural depression, 98
Natural disturbance, 5
Natural Sciences Society, 11
Nature (Chapman), 199
Nature (Farman), 187
Nature (Joseph, Farman, Gardiner &
        Shanklin), 1
Nature (Molina and Rowland), 169,
        170*f*, 175
Nature (Solomon), 197, 203
Naval Air Weapons Station, 129n6
Naval Ordnance Test Station (NOTS),
        129n6
Newman, Paul, 232
Newton's theory of gravity, 147
Niagara Falls, 25
Nicolet, Baron Marcel, 2*b*, 2*f*, 126, 127, 127*f*,
        128, 129*f*, 130, 130n7, 132
Nierenberg, William Aaron, 206*b*
Nimbus 4 satellite, 115
Nimbus 7 satellite, 115, 190*b*, 193
nitric oxide, 132, 219–220
    concentration, in situ
        measurements, 133
    oxidation of, 132
nitrogen, 7, 199n4
    ozone destruction by, 130–136
nitrogen dioxide, 29
nitrogen oxide, 7, 135
    ionization of, 127n3
    released by aircraft engines, 159–166

Rocket Research Center (Peenemünde), 93, 93*f*

rockets, ozone measurements from, 111–113

Rodhe, Henning, 131

Rossby, Carl-Gustaf, 144, 169

Rossby waves, 146*f*

Rouche, Nicolas, 230

Rowland, F. Sherwood, 169, 170, 170*f*, 178*f*, 197

Rowland, Sherry, 170, 171, 171*f*, 175, 175*f*, 176, 190*b*, 202, 231*f*

Royal Air Force, 93, 140

Rubin, Mordecai B., 20*b*

Ruderman, Melvin A., 160

Runge Tolme, Carl David, 77, 78*f*, 124

Russell, James, 115, 116

SAGE. *See* Stratospheric Aerosol and Gas Experiment (SAGE)

Sainte-Claire Deville, Charles Joseph, 34, 35*f*

satellite measurements, 113–118

SBUV. *See* Solar Backscatter Ultraviolet Radiometer (SBUV)

SCEP. *See* Study of Critical Environmental Problems (SCEP)

Schauerhammer, Ralf, 206*b*

Scherhag, Richard, 145

Schiff, Harold, 133, 164*b*

Schonbein, C., 12

Schönbein, Friedrich, 5, 9, 10, 10*f*, 11, 12, 12n5, 13–15, 16*b*, 19, 20, 25, 27, 28, 28*b*, 30n3, 32, 39, 44

Schönbein scale, 28

Schumann, Victor, 77, 78*f*, 124

Scottish Meteorological Society, 30

seasonal evolution, of ozone, 34*f*

seasonal variation, 119
of ozone column, 58*f*

Second World War, US Weather Bureau and during, 169

Seitz, Fred, 206*b*

semidiurnal tides, 145

Shanklin, Jonathan, 184, 186*f*

Shimazaki, Tatsuo, 148

short-wave radiation, 217

Singer, S. Fred, 2*b*, 205, 206*b*

skepticism, 185

smell of ozone, 12*b*

smog
chemical reactions, 220*b*
formation, automobiles in, 219
London-type, 214, 216*f*
Los Angeles-type, 213

smoke concentration, 215*f*

Solar Backscatter Ultraviolet Radiometer (SBUV), 115, 188, 190*b*, 194, 224*f*

solar light, 227

solar radiation, 30*f*, 50, 64*b*, 76, 98, 124, 221
Earth's surface of, 54
intensity of, 85, 124
station, 65
ultraviolet, 76
at wavelength, 51–52

solar spectrum, 121
at different altitudes, 112*f*

solar ultraviolet, 125
radiation, 53, 119

Solomon, Susan, 148, 197, 197*f*, 198, 200, 200*f*, 201–203

Soret, Jacques-Louis, 17–19*f*

southern hemisphere, 210n8
orography of, 144

Soviet-American nuclear explosions, 160

spacecraft, 117*f*

space measurements, of ozone, 118

space observations, of ozone depletion, 188–195

Space Shuttle Discovery, 117*f*

SPARC project. *See* Stratosphere-troposphere processes and their role in climate (SPARC) project

spectrometer, 52
microwave, 200

Spectrometer for Atmospheric Chartography (SCIAMACHY), 118